FUNDAMENTALS OF MANUFACTURING

Prepared by

Manufacturing Engineering Certification Institute of SME

Published by

Society of Manufacturing Engineers
One SME Drive
P.O. Box 930
Dearborn, Michigan 48121-0930

FUNDAMENTALS OF MANUFACTURING

Library of Congress Catalog Number 93-086050
International Standard Book Number 087263-446-9
Manufactured in the United States of America

Preface

This book was designed to provide a structured review for the individual planning to take the Fundamentals of Manufacturing Certification Examination. The topics covered are the result of a 1989 Delphi study based on a survey of manufacturing managers, manufacturing technologists and engineers, and manufacturing educators. Its purpose: to identify fundamental competency areas required by manufacturing technologists and entry-level engineers in the field.

While the objective of this book is to help prepare manufacturing managers, engineers, and technologists for the certification process, it is also a primary source of information for manufacturing and may serve as a reference for a variety of other purposes.

Acknowledgements
The Certification Committee of SME's Manufacturing Engineering Certification Institute expresses its sincere gratitude to a host of manufacturing professionals whose contributions have made this book possible.

Authoring major sections of the text were Dr. Paul Eagle, CMfgE, PE; Mr. Jean Younes; Dr. Tracey Tillman, CMfgE; Mr. Philip Rufe; Dr. Max Kanagy, CMfgE; and Dr. Walter Tucker.

Lending their expertise and critical assessment in the form of technical review were Dr. Jeffrey Abell, CMfgE; Dr. Paul Plotkowski, CMfgE; Dr. Tracey Tillman; and Mr. Warren Worthley, CMfgE.

The Committee, the Institute, and SME is deeply indebted to these individuals for their tireless efforts and commitment to excellence in the compilation and preparation of this book.

CONTENTS

20. Management and Economics

21. Computer Applications/Automation

INTRODUCTION

Manufacturing Certification

Manufacturing is concerned with energy, materials, tools, equipment, and products. Excluding services and raw materials in their natural state, most of the remaining gross national product is a direct result of manufacturing.

Modern manufacturing activities have become exceedingly complex because of rapidly increasing technology and an expanded environmental involvement. This, coupled with increasing social, political, and economic pressures, has caused successful firms to strive for high-quality manufacturing engineers and managers.

To aid in improving manufacturing, the Society of Manufacturing Engineers' Board of Directors initiated the Certification Program in 1972. The principal advantage of certification is that it shows ability to meet a certain set of standards related to the many aspects of manufacturing. These standards pertain to the minimum academic requirements needed, but more importantly, they pertain to the practical experience required of a manufacturing engineer or manager.

Many persons currently employed in industry can successfully measure themselves against these standards, but they cannot provide documentation concerning their ability. The certification program is designed to provide successful candidates with documentary evidence of their abilities.

Philosophically, the purpose of Manufacturing Certification is to gain increased acceptance of Manufacturing Engineering and Management as a profession and to ultimately improve overall manufacturing effectiveness and productivity.

Purpose and Overview

The purpose of this text is to provide a structured review for the manufacturing engineer or manager wishing to prepare for the Certification examination. The text will serve as a useful review for the experienced manufacturing professional or as an information source and study guideline for the person just entering the field.

The major areas of manufacturing science reviewed include mathematics, physics, materials science and metallurgy, engineering drawing, planning and strategy, and metrics and the SI system. In each area the topics emphasized are those fundamental to basic manufacturing processes.

It is assumed that the reader has completed some prior study of manufacturing science topics and has an experience-based working knowledge of tooling and manufacturing equipment. In light of these assumptions, certain basic science areas have not been included in the text, e.g., chemistry. For more detailed information on these topics, the reader is encouraged to consult the appropriate sources listed in the bibliography.

Sample problems and questions are included at the conclusion of each section to provide practice. Answers are included for all questions but it is recommended that all problems be attempted before reading the solution.

Part 1

Mathematical Fundamentals

MATHEMATICS

All aspects of engineering require the use of mathematics to analyze and design physical systems. This section provides a brief review of the basic concepts in algebra, geometry, trigonometry, probability, statistics and calculus. The material presented here is not complete and is not intended to be a resource for learning these topics for the first time.

1.1 Algebra

The study of algebra involves examining the basic properties of numbers. Algebra is founded on several basic laws. These laws or *axioms* can be used to derive all other concepts in algebra. Two of these basic laws apply to both addition and multiplication:

commutative law:
$$a + b = b + a$$
$$ab = ba$$

associative law:
$$a + (b + c) = (a + b) + c$$
$$a(bc) = (ab)c$$

Certain rules apply when handling exponents in algebra problems. For positive valued x and y the following rules apply:

$$x^{-a} = \frac{1}{x^a}$$

$$x^a x^b = x^{a+b}$$

$$(xy)^a = x^a y^a$$

$$x^{ab} = (x^a)^b$$

Logarithms are closely related to exponents. A *logarithm* is an exponent to which a base number is raised to give a particular result. If:

$$x = \log_b y$$

then x is the base b logarithm of y or

$$y = b^x$$

Engineering applications frequently use two types of logarithms: common or base 10 logarithms and so-called natural logarithms having base e (where $e = 2.7183...$). If x is the natural logarithm of y, then it can be written as:

$$x = \log_e y = \ln y$$

The logarithm of a negative number does not exist. The logarithm of numbers less than one are negative. The logarithm of one is zero. The logarithm of numbers greater than one are positive. There are several general rules useful in solving problems involving logarithms. These rules apply to logarithms of any base.

$$\log(xy) = \log x + \log y$$

$$\log x^a = a \log x$$

$$\log\left(\frac{x}{y}\right) = \log x - \log y$$

$$\log_b b = 1$$

$$\log 1 = 0$$

Example 1.1.1. If $\log_a 10 = 0.25$ then what is $\log_{10} a$?
Solution. The expression $\log_a 10 = 0.25$ can be rewritten as $10 = a^{0.25}$. The base 10 logarithm can be taken of both sides of the expression to give:
$$\log_{10} 10 = \log_{10} a^{0.25}$$

$$1 = 0.25 \log_{10} a$$

or

$$\log_{10}a = \frac{1}{0.25} = 4$$

Example 1.1.2. The logarithm of 2 in base 10 is known to be 0.30103. Find the logarithm to the base 10 of 0.5.
Solution. The expression $\log_{10}\frac{1}{2}$ can be rewritten as

$$\log_{10}\frac{1}{2} = \log_{10}1 - \log_{10}2 = 0.0 - 0.30103 = -0.30103$$

One of the most important applications of algebra is solving equations with one variable or unknown. The most commonly used forms are: linear equations and quadratic equations. Linear equations with one unknown have the basic form:

$$ax + b = 0$$

The unknown quantity x can be solved for by successive application of the following rules.

An equivalent equation can be made by:

• adding the same number to, or subtracting the same number from, both sides of an equation.

• multiplying or dividing both sides of an equation by the same number.

Another form of an equation of a single variable is the quadratic equation. The basic form of a quadratic equation is given by:

$$ax^2 + bx + c = 0$$

There is a standard solution to a quadratic equation:

$$x = \frac{-b \pm \sqrt{b^2 - 4ac}}{2a}$$

Note that there may be *two* possible solutions. The number of solutions and their type depends on the value of the *discriminant* (the quantity under the radical). If $b^2 - 4ac > 0$, the quadratic equation has two distinct (different), real solutions. If $b^2 - 4ac = 0$, then the equation has one real solution. If $b^2 - 4ac < 0$, then the equation has two distinct, imaginary (involving i or $\sqrt{-1}$) solutions.

Example 1.1.3. Find the value of x.

$$\frac{4x+2}{x+1} + \frac{4}{5} = -\frac{6}{5}$$

Solution. The lowest common denominator is $5(x+1)$. Both sides of the equation are multiplied by it to obtain:

$$20x+10+4x+4 = -6x-6$$

$$20x+4x+6x = -6-4-10$$

$$30x = -20$$

$$x = -\frac{2}{3}$$

Example 1.1.4. Find the solution(s) to the following equation.

$$17x^2+41x-74 = 19$$

Solution. The equation is first put in standard form:

$$17x^2+41x-93 = 0$$

where $a = 17$, $b = 41$ and $c = -93$

Since the discriminant is greater than zero:

$$b^2-4ac = (41)^2-(4)(17)(-93)>0$$

there are two, real solutions.

$$x_{1,2} = \frac{-41\pm\sqrt{41^2-4(17)(-93)}}{(2)(17)} = -3.84, \ 1.42$$

A linear equation with two unknowns has the general form given by:

$$ax + by = c$$

where this is an equation of a line with slope $-a/b$ and y-intercept c/b as shown in Figure 1.1. Two *independent* equations are needed to find a unique solution (the two equations cannot simply be multiples of each other). A unique solution represents the point where the two lines intersect. There will be no unique solution if the lines are parallel (the slopes are identical).

There are three major approaches to solving these linear, simultaneous systems of equations:

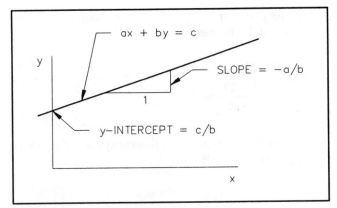

Figure 1.1 Equation of a Line.

• **Substitution.** Substitution involves solving for one variable in terms of the other and substituting the result in the other equation.

• **Elimination.** Elimination can be used to solve systems of linear simultaneous equations by multiplying both equations by an appropriate constant such that one of the unknown terms cancels out when the two equations are added together.

• **Cramer's Rule.** Cramer's Rule uses *determinants*, a special function of the coefficients in the equations, to solve for the variables.

Example 1.1.5. Solve the following system of two linear simultaneous equations using substitution.

$$2x+3y = 1$$
$$3x-5y = -27$$

Solution. Solve the first equation for y in terms of x:

$$y = -\frac{2}{3}x + \frac{1}{3}$$

Substituting, the equation becomes:

$$3x-5(-\frac{2}{3}x+\frac{1}{3}) = -27$$

Clearing fractions and the parenthesis results in the solution for x:

$$9x+10x-5 = -81$$
$$19x = -76$$
$$x = -4$$

The resulting value of x is substituted in either equation to obtain the solution for y:

$$2x+3y = 1$$
$$-8+3y = 1$$
$$3y = 9$$
$$y = 3$$

Example 1.1.6. Solve the system of equations in example 1.1.5 by elimination.

Solution. Multiply the first equation by five and divide the second equation by three to obtain y terms which will cancel each other out:

$$10x+15y = 5$$
$$9x-15y = -81$$
$$\overline{}$$
$$19x = -76$$
$$x = 4$$

The result is substituted into either equation to obtain a solution for y as in example 1.1.5.

A third approach to solving linear, simultaneous equations is Cramer's Rule. This method uses *determinants* to find the unknowns. A determinant is a *scalar* (a one dimensional quantity) value that can be obtained by performing a special sequence of algebraic operations on a group of scalars. A 2x2 (read "2 by 2") determinant is written as:

$$\begin{vmatrix} a_1 & b_1 \\ a_2 & b_2 \end{vmatrix}$$

the determinant is evaluated as:

$$\begin{vmatrix} a_1 & b_1 \\ a_2 & b_2 \end{vmatrix} = a_1 b_2 - a_2 b_1$$

Cramer's Rule requires the simultaneous equations to be set up in the following standard form:

$$a_1 x + b_1 y = c_1$$
$$a_2 x + b_2 y = c_2$$

7

A determinant is formed for the coefficients as:

$$\det_{COEFF} = \begin{vmatrix} a_1 & b_1 \\ a_2 & b_2 \end{vmatrix}$$

A modified determinant for x is formed by substituting the constants (the terms on the right side of the equation in standard form) for the coefficients of x:

$$\det_x = \begin{vmatrix} c_1 & b_1 \\ c_2 & b_2 \end{vmatrix}$$

A modified determinant for y is formed by substituting the constants for the coefficients of y:

$$\det_y = \begin{vmatrix} a_1 & c_1 \\ a_2 & c_2 \end{vmatrix}$$

The solutions for x and y can be found by dividing the appropriate modified determinant by the coefficient determinant:

$$x = \frac{\det_x}{\det_{COEFF}} = \frac{\begin{vmatrix} c_1 & b_1 \\ c_2 & b_2 \end{vmatrix}}{\begin{vmatrix} a_1 & b_1 \\ a_2 & b_2 \end{vmatrix}}$$

and

$$y = \frac{\det_y}{\det_{COEFF}} = \frac{\begin{vmatrix} a_1 & c_1 \\ a_2 & c_2 \end{vmatrix}}{\begin{vmatrix} a_1 & b_1 \\ a_2 & b_2 \end{vmatrix}}$$

Example 1.1.7. Solve the system of equations in example 1.1.5 using Cramer's Rule.
Solution. Using determinants, x can be found as:

$$x = \frac{\begin{vmatrix} 1 & 3 \\ -27 & -5 \end{vmatrix}}{\begin{vmatrix} 2 & 3 \\ 3 & -5 \end{vmatrix}} = \frac{(1)(-5)-(-27)(3)}{(2)(-5)-(3)(3)} = \frac{76}{-19} = -4$$

The solution for y can be found easily since the coefficient determinant is known:

$$y = \frac{\begin{vmatrix} 2 & 1 \\ 3 & 27 \end{vmatrix}}{-19} = \frac{(2)(-27)-(3)(1)}{-19} = \frac{-57}{-19} = 3$$

Example 1.1.8. Solve the following system of equations using Cramer's Rule:

$$3x+4y = 2$$
$$9x+12y = 6$$

Solution. A solution for x is attempted by using determinants:

$$x = \frac{\begin{vmatrix} 2 & 4 \\ 6 & 12 \end{vmatrix}}{\begin{vmatrix} 3 & 4 \\ 9 & 12 \end{vmatrix}} = \frac{24-24}{36-36} = \frac{0}{0}$$

The result is an indeterminant form (division by zero is not allowed). Closer inspection of the two equations reveals that they are not independent. No solution exists because the lines represented by the two equations are parallel.

1.2 Geometry

The areas and volumes of common geometric shapes are commonly needed in the solution of engineering problems. The areas of some common two-dimensional shapes are shown in Figure 1.2. The volumes of some common three-dimensional shapes are given in Figure 1.3.

The equation of a straight line can be written in a variety of forms. The general form of an equation of a line is given by:

$$Ax + By + C = 0$$

Various other forms of the equation of a line are also used. The *point-slope* form is given by:

$$y - y_1 = m(x - x_1)$$

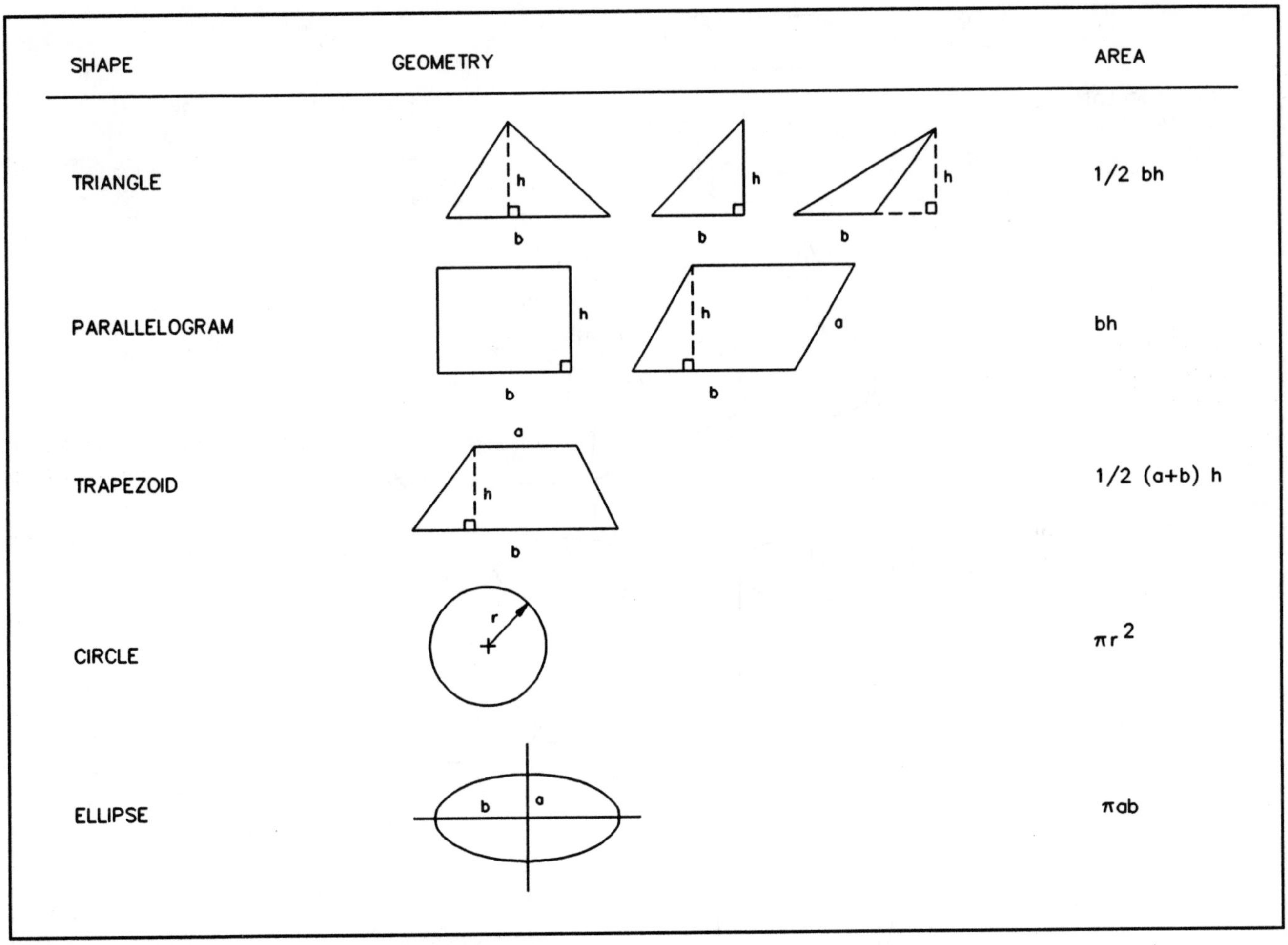

SHAPE	GEOMETRY	AREA
TRIANGLE		$1/2\ bh$
PARALLELOGRAM		bh
TRAPEZOID		$1/2\ (a+b)\ h$
CIRCLE		πr^2
ELLIPSE		πab

Figure 1.2 Areas of Various Two-dimensional Shapes.

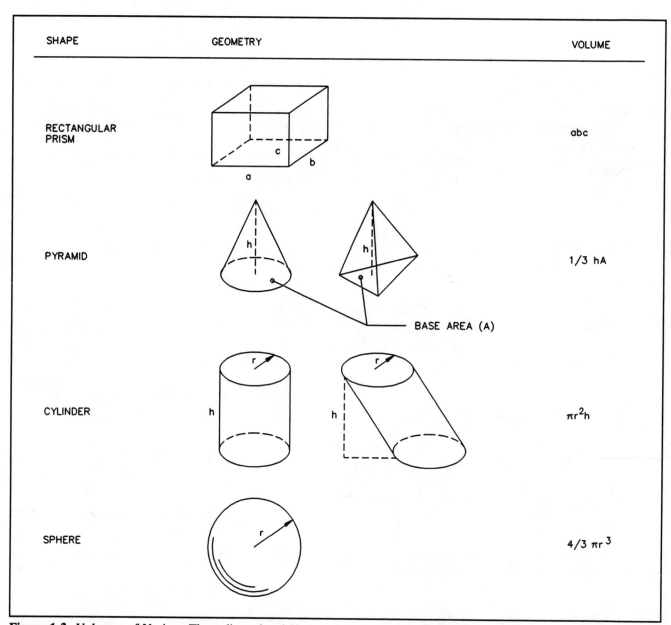

SHAPE	GEOMETRY	VOLUME
RECTANGULAR PRISM		abc
PYRAMID		$1/3\ hA$
CYLINDER		$\pi r^2 h$
SPHERE		$4/3\ \pi r^3$

Figure 1.3 Volumes of Various Three-dimensional Shapes.

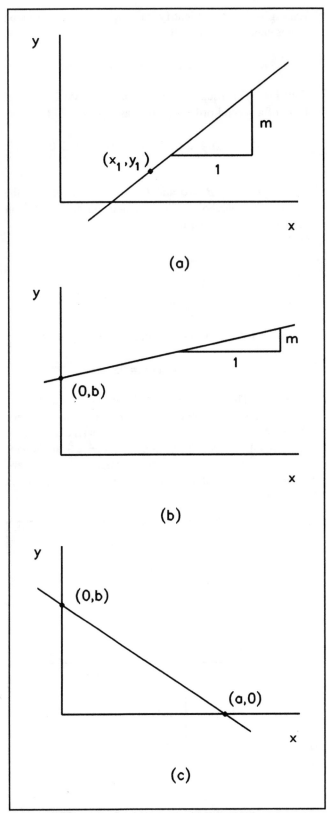

Figure 1.4 Various Forms of the Equation of a Line.

describing the line shown in Figure 1.4(a). The *slope-intercept* form is given by:

$$y = mx + b$$

describing the line shown in Figure 1.4(b). The *two-intercept* form is given by:

$$\frac{x}{a} + \frac{y}{b} = 1$$

describing the line shown in Figure 1.4(c).

Example 1.2.1. Find the equation describing the line shown in Figure 1.5.

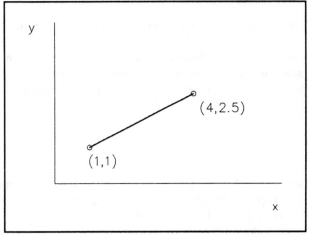

Figure 1.5 Line Described by Two Known Points.

Solution. The slope of the line is found as:

$$m = \frac{2.5-1}{4-1} = \frac{1}{2}$$

Either given point can be used in the point-slope form to give the equation of the line:

$$y - y_1 = m(x - x_1)$$
$$y - 1 = \frac{1}{2}(x - 1)$$
$$-2x + y + 1 = 0$$

Lines are *first degree* equations because the power of x in the equation is one. Circles, ellipses, parabolas and hyperbolas are described by *second degree* equations. The general form of a second degree equation is given by:

$$Ax^2 + Bxy + Cy^2 + Dx + Ey + F = 0$$

11

This equation describes a set of geometric shapes called *conic sections*. There are various forms of conic sections that can be categorized according to the magnitude of the coefficients A, B, C, D, E and F:

Circle	$B = 0$, $A = C$ and $B^2 - 4AC$ ·
Ellipse	$B^2 - 4AC < 0$
Parabola	$B^2 - 4AC = 0$
Hyperbola	$B^2 - 4AC > 0$

A *circle* is the locus or collection of points that are equidistant from a given point in a plane. The equation for a circle can be expressed in a general form as:

$$(x-h)^2 + (y-k)^2 = r^2$$

with center at (h,k) and radius r as shown in Figure 1.7(a).

An *ellipse* is the locus of points in a plane, the sum of whose distances from two fixed points (focal points or foci) is constant. The general equation for an ellipse centered at the origin is:

$$\frac{x^2}{a^2} + \frac{y^2}{b^2} = 1$$

where a and b are half the major and minor axis lengths. The foci are located at $(\pm c, 0)$ where $c^2 = a^2 - b^2$ as shown in Figure 1.7(b).

A *parabola* is the locus of points in a plane that are equidistant from a fixed focal point and a line known as the *directrix*. For a parabola with the vertex at (h,k), the general equation for a parabola is given by:

$$(x-h)^2 = 4p(y-k)$$

where the equation of the directrix line is $y = k-p$ and the focal point location is $(h, k+p)$.

A *hyperbola* is the locus of points in a plane, the difference of whose distances from two fixed points (focal points or foci) is constant. The general equation describing a hyperbola centered at the origin opening to the left and right is:

$$\frac{x^2}{a^2} - \frac{y^2}{b^2} = 1$$

The foci are given by $(\pm c, 0)$ where $c^2 = a^2 + b^2$. The hyperbola approaches two asymptotes defined by the lines $y = \pm \frac{b}{a} x$.

Example 1.2.2. Identify the type of conic section represented by the equation:

$$y = 2x^2 - 6x + 4$$

Solution. This equation can be converted into the general form of a second order equation as:

$$-2x^2 + 0xy + 0y^2 + 6x + 0y - 4 = 0$$

where $A = -2$, $B = 0$ and $C = 0$. Therefore, the equation is a parabola. The equation can be rewritten in the more recognizable form:

$$\left(x - \frac{3}{2}\right)^2 = \frac{1}{2}\left(y + \frac{1}{2}\right)$$

where $h = \frac{3}{2}$, $k = -\frac{1}{2}$, and $4p = \frac{1}{2}$, or $p = \frac{1}{8}$.

Example 1.2.3. Sketch the graph of the equation:

$$9x^2 + 4y^2 = 25$$

Solution. The coefficients of the general second order equation are $A = 9$ and $C = 4$ for this equation. Since $B^2 - 4AC = -36$, the equation describes an ellipse. If y is set equal to zero, the x-intercepts $x = \pm \frac{5}{3}$ are found. If x is set equal to the zero, the y-intercepts $y = \pm \frac{5}{2}$ are found. A sketch of the graph of the ellipse is given in Figure 1.6.

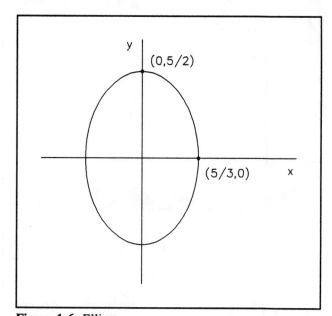

Figure 1.6 Ellipse.

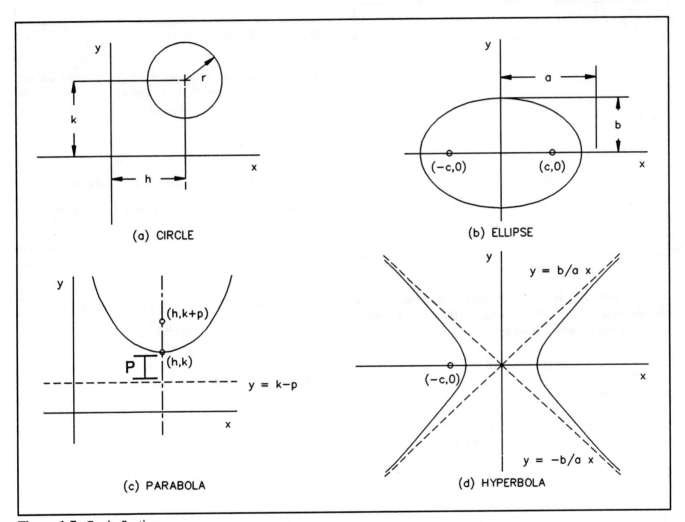

Figure 1.7 Conic Sections.

13

1.3 Trigonometry

The basic trigonometric functions can be easily defined as ratios of the sides of a right triangle. Using the right triangle shown in Figure 1.8, the following functions can be defined:

$$\sin\theta = \frac{y}{h} \quad \cos\theta = \frac{x}{h} \quad \tan\theta = \frac{y}{x}$$

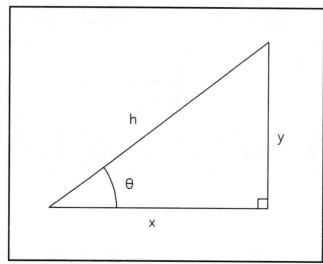

Figure 1.8 Right Triangle.

where h is known as the hypotenuse of the triangle. By application of the Pythagorean Theorem, a basic trigonometric identity can be found:

$$\cos^2\theta + \sin^2\theta = 1$$

for any angle θ.

There are three other trigonometric functions that can also be defined as reciprocals of the three basic functions:

$$\cot\theta = \frac{1}{\tan\theta} \qquad \sec\theta = \frac{1}{\cos\theta} \qquad \csc\theta = \frac{1}{\sin\theta}$$

The angle between the adjacent side and hypotenuse can be described in either radians or degrees. Degrees can be converted to radians by multiplying by $\pi/180°$. Another important observation is that angles are measured counterclockwise positive.

The tangent (tan) function asymptotically approaches infinity at odd multiples of $\pi/2$. The sine (sin) and cosine (cos) functions are periodic in 2π intervals and are out of phase with each other by $\pi/2$ or $90°$.

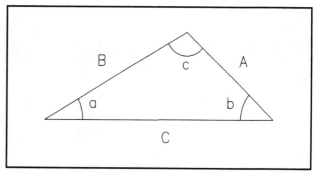

Figure 1.9 General Triangle.

The following is a list of a number of useful trigonometric identities that can be derived from the definitions of the trigonometric functions and the Pythagorean identity. (Note: u and v are any angle.)

Negative Angle Formulas

$$\sin(-u) = -\sin u \quad \cos(-u) = \cos u \quad \tan(-u) = -\tan u$$

$$\csc(-u) = -\csc u \quad \sec(-u) = \sec u \quad \cot(-u) = -\cot u$$

Addition Formulas

$$\sin(u\pm v) = \sin u \cos v \pm \cos u \sin v$$

$$\cos(u\pm v) = \cos u \cos v \mp \sin u \sin v$$

$$\tan(u\pm v) = \frac{\tan u \pm \tan v}{1 \mp \tan u \, \tan v}$$

Double Angle Formulas

$$\sin 2u = 2 \sin u \cos u$$

$$\cos 2u = \cos^2 u - \sin^2 u = 1 - 2\sin^2 u = 2\cos^2 u - 1$$

$$\tan 2u = \frac{2 \tan u}{1 - \tan^2 u}$$

Half Angle Formulas

$$\sin^2\frac{u}{2} = \frac{1 - \cos u}{2}$$

$$\cos^2\frac{u}{2} = \frac{1 + \cos u}{2}$$

$$\tan\frac{u}{2} = \frac{1 - \cos u}{\sin u} = \frac{\sin u}{1 + \cos u}$$

There are two important formulas that apply to a general triangle shown in Figure 1.9. The *law of sines* describes the relationship of the angles in a triangle to the sides opposite them:

$$\frac{\sin a}{A} = \frac{\sin b}{B} = \frac{\sin c}{C}$$

where $a + b + c = 180°$. The *law of cosines* describes the relationships between the sides of a triangle and an angle:

$$A^2 = B^2 + C^2 - 2BC \cos a$$

Example 1.3.1. If a particular equilateral triangle has a side with a length of 1, what is the distance from the side to its center?

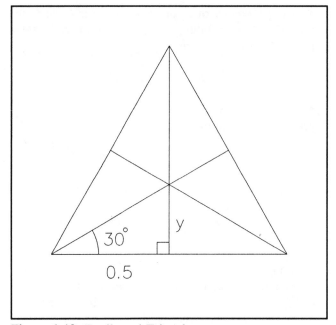

Figure 1.10 Equilateral Triangle.

Solution. Referring to Figure 1.10, the equilateral triangle can be divided into six smaller triangles. Applying the definition of the tan function:

$$\tan 30° = \frac{y}{0.5}$$

$$y = 0.29$$

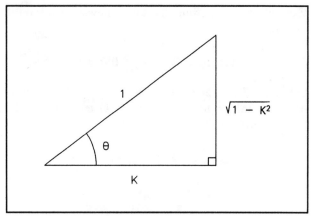

Figure 1.11 Right Triangle.

Example 1.3.2. If $\cos \theta = k$, what is $\tan \theta$?
Solution. If a right triangle is drawn as in Figure 1.11, $\cos \theta = k$ implies that the hypotenuse of the triangle is 1 since $\cos \theta = x/h = k/1 = k$. The Pythagorean Theorem can be applied to determine the length of the other side as $\sqrt{1 - k^2}$. The value of the **tan** function can be found as:

$$\tan \theta = \frac{\sqrt{1 - k^2}}{k}$$

Example 1.3.3. If a triangle has two sides of length 5.3 and 3.8 with a 110° angle between them, what is the length of the third side?
Solution. The triangle is shown in Figure 1.12. The law of cosines can be applied directly as:

$$A^2 = 3.8^2 + 5.3^2 - 2(3.8)(5.3) \cos 110°$$

$$A = 7.5$$

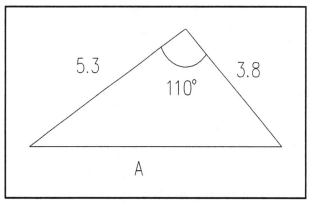

Figure 1.12 Application of the Law of Cosines.

Example 1.3.4. Simplify the following trigonometric expression:

$$\tan \theta(\sec \theta(1 - \sin^2 \theta)/\cos \theta)$$

Solution. The expression can be simplified by substitution of the definition for each function in terms of the basic trigonometric functions.

$$\tan \theta \; \sec \theta \; (1 - \sin^2 \theta) / \cos \theta$$

$$= \frac{\sin \theta}{\cos \theta} \cdot \frac{1}{\cos \theta} \cdot \frac{\cos^2 \theta}{\cos \theta}$$

$$= \tan \theta$$

Example 1.3.5. Find the unknown side x in the triangle in Figure 1.13.

Solution. The angle α can be found since the interior angles in a triangle must sum to $180°$.

$$180° = \alpha + 35° + 45°$$

$$\alpha = 100°$$

The law of sines can be applied directly to find the unknown side:

$$\frac{\sin 100°}{x} = \frac{\sin 45°}{20}$$

$$\frac{0.985}{x} = \frac{0.707}{20}$$

$$x = 27.86$$

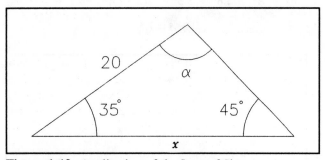

Figure 1.13 Application of the Law of Sines.

1.4 Probability

A *permutation* of any number of items is a group of some or all of them arranged in a definite order. The number of permutations of n things taken r at a time is designated $p(n,r)$ and can be determined by:

$$p(n,r) = \frac{n!}{(n - r)!}$$

where the exclamation point refers to the factorial (e.g., $4! = 1 \cdot 2 \cdot 3 \cdot 4 = 24$).

A *combination* in a group of n objects is any set of r objects taken without regard to their order. The number of combinations of n things taken r at a time is given by:

$$C(n,r) = \frac{n!}{r! \, (n - r)!}$$

A probability $P(A)$ refers to the likelihood of an event A occurring. The probability is always in the range of from zero to one. For two independent events, A and B, the following rules of probability apply:

- The probability of A or B occurring is equal to the sum of the probability of occurrence of A and the probability of occurrence of B.

$$P(A \; or \; B) = P(A) + P(B)$$

- The probability of A and B occurring is given by the product of the two probabilities.

$$P(A \; and \; B) = P(A) \, P(B)$$

- The probability of A not occurring is given by:

$$P(not \; A) = 1 - P(A)$$

If at any given trial, an event can occur in h different ways, fail to occur in f different ways and all of the h and f are equally likely, then the probability of the event occurring $P(h)$ at a given trial is:

$$P(h) = \frac{h}{h + f}$$

Example 1.4.1. There are seven ball bearings packed in a tube. How many different ways can the seven bearings be arranged?

Solution. This is determined by the number of permutations of seven things taken seven at a time:

$$p(7,7) = \frac{n!}{(n-r)!} = \frac{7!}{(7-7)!} = 5040$$

Example 1.4.2. How many different collections of 10 people can fill three available offices with one to an office?
Solution. The answer does not depend on ordering. It is a combination problem.

$$C(10,3) = \frac{n!}{(n-r)!r!} = \frac{10!}{(10-3)!3!} = 120$$

Example 1.4.3. An ordinary die is tossed 100 times. What is the expected number of times a four is rolled?
Solution. The probability of a four on any toss is $1/6$ and all events are independent. Let 4_n represent tossing a four on the nth toss.

$$P(4_1 \text{ and } 4_2 \text{ and } 4_3 \dots 4_{100}) = \frac{1}{6} \cdot \frac{1}{6} \cdot \frac{1}{6} \dots \frac{1}{6} = 100 \cdot \frac{1}{6} \approx 17$$

Approximately 17 fours would appear in 100 tosses.

Example 1.4.4. What is the probability of drawing a king or a one-eyed jack from an ordinary deck of 52 playing cards?
Solution. The events of drawing a king or one-eyed jack are independent. The probability of drawing a king is $4/52$. The probability of drawing a one-eyed jack is $2/52$. The probability of selecting a king or a one-eyed jack is:

$$P(king \text{ or } o.e. \text{ } jack) = P(king) + P(o.e. \text{ } jack)$$

$$= \frac{4}{52} + \frac{2}{52} = \frac{6}{52} = 0.115$$

1.5 Statistics

Statistics is the area of mathematics that describes the characteristics of a large system by use of parameters that characterize that system. Statistics are the quantities used to analyze data gathered in an experiment. A statistic is a single number that represent some characteristic of a group of data.

The *arithmetic mean* $\bar{x}$ is the expected value or average of a group of N observations.

$$\bar{x} = \frac{1}{N} \sum_{i=1}^{N} x_i$$

The *median* is the middle observation in a group of data ordered by magnitude. One half of the values are below the median. By definition, the median IQ is 100.

The *mode* is the value that occurs most frequently. The mode of the age of high school students is typically 16.

The *standard deviation* is a measure of variability in a set of data. It is defined as:

$$\sigma = \left[\frac{\sum_{i=1}^{N} (x_i - \bar{x})^2}{N} \right]^{1/2} = \left[\frac{\sum_{i=1}^{N} (x_i^2) - \left(\sum_{i=1}^{N} x_i\right)^2}{N} \right]^{1/2}$$

This is the population standard deviation (all of the available data is assumed to be used). If a small sample of available data is used (typically less than 50), use $n-1$ in the denominator above. The *variance* is defined as σ^2.

Example 1.5.1. Calculate the mean and standard deviation of 85, 70, 60, 90 and 81. Treat these numbers as samples drawn from a large population.
Solution. The mean is found as:

$$\bar{x} = \frac{386}{5} = 77.2$$

The sample standard deviation is found as:

$$\sigma = \left[\frac{\sum_{i=1}^{N} (x_i^2) - \left(\sum_{i=1}^{N} x_i\right)^2}{N-1} \right]^{1/2} = \left[\frac{30{,}386 - \frac{(386)^2}{5}}{4} \right]^{1/2} = 12.1$$

1.6 Calculus

A knowledge of calculus is the cornerstone of a sound engineering education. Calculus is briefly reviewed here in its most basic elements.

Differential calculus involves the examination of how something changes relative to something else. For example, if the position of an object is known as a function of time, then differentiation of that function (or the *derivative* of the function) will give the velocity. The slope of a function $y = f(x)$ at a point is the first derivative of the function written as:

$$\frac{dy}{dx} = Dy = y'$$

The second derivative of a function is written as:

$$\frac{d^2y}{dx^2} = D^2y = y''$$

Some derivative formulas are given below. The notation assumes that f and g are functions of x, and k is a constant.

$$\frac{dk}{dx} = 0$$

$$\frac{d(kx^n)}{dx} = kx^{n-1}$$

$$\frac{d}{dx}(f+g) = f' + g'$$

$$\frac{df^n}{dx} = nf^{n-1}f'$$

$$\frac{d}{dx}\left(\frac{f}{g}\right) = \frac{gf' - fg'}{g^2}$$

$$\frac{d}{dx}(fg) = fg' + f'g$$

$$\frac{d}{dx}(\ln x) = \frac{1}{x}$$

$$\frac{d}{dx}(e^{kx}) = ke^{kx}$$

$$\frac{d}{dx}(\sin x) = \cos x$$

$$\frac{d}{dx}(\cos x) = -\sin x$$

The slope of a function can be found at a point if the derivative of the function can be evaluated at that point.

Example 1.6.1. Find the slope of the function $y = x^3 - 4x$ at the origin.
Solution. The derivative can be evaluated as:

$$\frac{dy}{dx}\bigg|_{x=0} = [3x^2 - 4]_{x=0} = -4$$

Example 1.6.2. The position of a particle as a function of time is given as:

$$x = t^2 + \sin t$$

Find the velocity at $t = 0.5$.
Solution. The velocity is the derivative evaluated at the desired point:

$$velocity = \frac{dx}{dt} = [2t + \cos t]_{t=0.5} = 1.878$$

Example 1.6.3. Find the derivative of $\cot x$ with respect to x.
Solution. Writing $\cot x$ as $\cos x / \sin x = f(x)/g(x)$ it follows that:

$$\frac{(\sin x)(-\sin x) - (\cos x)(\cos x)}{\sin^2 x} = -\frac{1}{\sin^2 x} = -\csc^2 x$$

One of the most important applications for derivatives is finding the maximum and minimum points in a function $f(x)$. The following tests can be applied to determine the location of a maximum or minimum point:

$f'(x) = 0$ at a maximum or minimum point
$f''(x) > 0$ at a minimum point
$f''(x) < 0$ at a maximum point

Example 1.6.4. Find the maximum and minimum points in the function:

$$f(x) = x^3 - 3x^2 + 3$$

Solution. The derivative of the function is set equal to zero to reveal the extreme points in the function.

$$f'(x) = 3x^2 - 6x = 3x(x-2) = 0$$

$$3x = 0 \quad and \quad x-2 = 0$$

$$x = 0 \quad and \quad x = 2$$

The second derivative is used to test whether the points are maximum or minimum points:

$$f''(x) = 6x - 6$$

$$f''(0) = -6 \; (maximum)$$

$$f''(2) = 6 \; (minimum)$$

This is confirmed by the graph of the function in Figure 1.14.

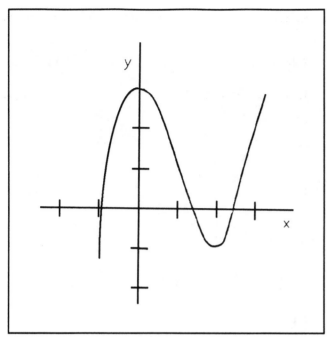

Figure 1.14 Plot of the Function in Example 1.6.4.

Integral calculus is the inverse of differentiation. If the derivative of a function is integrated, the original function plus a constant of integration is the result:

$$\int f'(x)\,dx = f(x) + C$$

This is referred to as an *indefinite integral*. A *definite integral* is one that is evaluated between two limits of integration. If a function describing a curve is given by $y = f(x)$, then the area under that curve from a to b is given by:

$$A = \int_a^b f(x)\,dx$$

as shown in Figure 1.15.

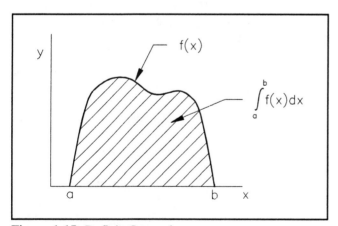

Figure 1.15 Definite Integral.

Some common integration formulas are given below:

$$\int dx = x$$

$$\int c\,f(x)\,dx = c\int f(x)\,dx$$

$$\int x^n\,dx = \frac{x^{n+1}}{n+1} + C \quad n \neq -1$$

$$\int x^{-1}\,dx = \ln x + C$$

$$\int e^{ax}\,dx = \frac{1}{a}e^{ax} + C$$

$$\int \sin x\,dx = -\cos x + C$$

$$\int \cos x\,dx = \sin x + C$$

Example 1.6.5. Evaluate the indefinite integral:

$$\int (x^3 + 1)^2\,dx$$

Solution. Squaring the quantity being integrated gives:

$$\int (x^3 + 1)^2\,dx = \int (x^6 + 2x^3 + 1)\,dx = \frac{1}{7}x^7 + \frac{2}{4}x^4 + x$$

Example 1.6.6. Find the area between the x axis and the function $y = x^2$ from 1 to 7.

Solution. The area can be found by application of the definite integral.

$$A = \int_1^7 x^2\,dx = \frac{x^3}{3}\bigg|_1^7 = \frac{343}{3} - \frac{1}{3} = \frac{342}{3}$$

19

Practice Problems

Solve the equations in Problems 1.1 - 1.4 for x:

1.1 $3x + 2 = 4x - 6 + x$

1.2 $\dfrac{1}{6}x - \dfrac{2}{3} = \dfrac{3}{4}x + \dfrac{1}{2}$

1.3 $\dfrac{x}{y} = 2x + 3$

1.4 $\dfrac{3}{x} - \dfrac{2}{x-1} = \dfrac{1}{2x}$

1.5   The length of a rectangle is four feet less than two times its width. The perimeter is 46 feet. What is the width?

1.6   A student must have an average grade from 80% to 90% on four tests to receive a grade of B. Grades on the first three tests were 83%, 76%, and 79%. What grade on the fourth test will guarantee a grade of B?

Solve the equations in Problems 1.7 - 1.10 by the quadratic theorem:

1.7 $y^2 + 5y = -5$

1.8 $(g + 2)(g - 3) = 1$

1.9 $6r^2 = 7 - 19r$

1.10 $\sqrt{2x + 3} = 2 + \sqrt{x - 2}$

1.11   A square has an area of 64 square inches. If the same amount is removed from one dimension and added to the other, the resulting rectangle has an area nine square inches less. Find the dimensions of the rectangle.

1.12   A pool is 30 feet by 40 feet. Wood chips are to be spread in a uniform width around the pool. If there are enough wood chips to cover 296 square feet, how wide can the strip be?

1.13   A rectangular piece of sheet metal has a length that is four inches less than its width. A piece two inches square is cut from each corner. The sides are turned up to form an uncovered box of volume 256 cubic inches. Find the dimensions of the original piece of metal.

Solve the systems of equations in Problems 1.14 - 1.16:

1.14 $2x - 3y = 10$
            $2x + 2y = 5$

1.15 $3x - 2y = 5$
            $-4x + 5y = 5$

1.16 $x + \dfrac{y}{3} = \dfrac{7}{3}$
            $x - \dfrac{5}{2}y = -\dfrac{1}{2}$

Solve the equations in Problems 1.17 - 1.19:

1.17 $25^x = 125$

1.18 $\log_2 x = 3$

1.19 $\log_{10} x + \log_{10} \dfrac{3x}{2} = 5$

1.20   A 10-meter-diameter cylindrical tank that sits on the ground is to be painted. The tank is 10 meters high. If one liter of paint covers five square meters, how many liters are required? Include the top.

1.21   A line with slope of -2 intersects the x axis at 2. Find the equation of the line.

1.22   A line intercepts the x axis at 4 and the y axis at -6. Find the equation of the line.

1.23   A track is 150 meters long and 75 meters wide. It has an elliptical shape. Find the equation of the ellipse.

1.24   Find the equation of a circle centered at (5,8) with a radius of eight units.

1.25   The asymptotes of a hyperbola are at ±45° with respect to the x. The hyperbola passes through the origin. Find the equation of the hyperbola.

1.26   What type of conic section is represented by the equation $x^2 + 4xy + 4y^2 + 2x - 10$?

1.27   If $\cos \theta = 0.8$, what is θ?

1.28 Two legs of a right triangle are three and four units long. How long is the hypotenuse?

1.29 Find the angle opposite the long leg of the triangle in Problem 1.27.

1.30 Find the interior angles of a triangle with sides two, three, and four.

1.31 Simplify the following trigonometric expression:
$$\tan^2\theta \ \csc^2\theta \ \sin2\theta$$

1.32 Express $\cos 2\theta$ as a function of $\sin^2\theta$.

1.33 A triangle has two sides of length six and eight units. The angle between these two sides is 60 degrees. Find the length of the side opposite the 60-degree angle.

1.34 A card is drawn from a well-shuffled deck of 52 cards. Find the probability that the card is:
a. a seven
b. a red seven
c. a spade
d. the seven of spades
e. a face card

1.35 Two dice are rolled. Find the probability that:
a. a seven is rolled
b. a number greater than two is rolled
c. an odd number is rolled

1.36 How many ways can six people be seated in a row of six seats?

1.37 How many different license plate numbers can be formed using three letters followed by three digits if no repeats are allowed?

1.38 A club has 30 members. If a committee of three is selected in a random manner, how many committees are possible?

1.39 How many different 2-card hands can be drawn from a 52-card deck?

1.40 One card is drawn from a standard deck of 52 cards. What is the probability that the card is:
a. a nine or ten
b. a diamond or a black

1.41 Two coins are tossed. Find the probability of both heads or both tails.

1.42 A coin is tossed three times. Find the probability of:
a. All three tosses resulting in tails
b. Only the first two tosses are tails

1.43 A teacher gives the following scores on an examination:

Frequency	Score
1	35
3	45
6	55
11	65
13	75
10	85
2	95

a. What is the mode of the scores?
b. What is the mean?
c. What is the standard deviation?

Differentiate the expressions in Problems 1.44 - 1.46 with respect to x:

1.44 $3x^5 + 7x^2$

1.45 $(\sin x)\,x^3$

1.46 $\dfrac{x^2-4}{3x}$

1.47 Find the slope of the function $y = 2x^3 - 3x$ at $x = 1$.

1.48 At what value of x does the minimum value of $y = x^3 - 3x$ occur?

1.49 Find the area under the curve $y=4x^3$ from $x = 1$ to $x = 2$.

1.50 Evaluate the integral $\displaystyle\int_0^\pi (e^x + \sin x)\,dx$

VECTOR

MECHANICS

2.1 Scalars and Vectors

There are two types of physical quantities, *scalars* and *vectors*. Scalar quantities are defined by magnitude alone. Some examples of scalar quantities are mass, speed, displacement, and power. The conventional algebraic operations used on real numbers are also used on scalar quantities. Vectors represent physical quantities that have both magnitude and direction. Some examples of vector quantities are velocity, acceleration and force. Vectors can be combined according to certain specialized algebraic rules. Scalars will be shown in this text with normal type, such as A. Vectors will be shown in bold, such as B. Other common notations are $\vec{B}$ and $\bar{B}$.

2.2 Addition of Vectors

Vectors can be more rigorously defined as *quantities possessing magnitude and direction which add according to the parallelogram law.* Two vectors, P and Q, can be summed by attaching the "tail" of each vector together and constructing a parallelogram with sides P and Q. The

diagonal represents the sum of the two original vectors as shown in Figure 2.1. Vector addition follows the commutative and associative properties.

The negative of a given vector P is the vector that has the same magnitude and direction, but opposite sense as in Figure 2.2. The negative of P is denoted as $-P$.

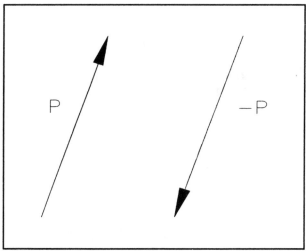

Figure 2.2 Negative of a Vector.

2.3 Rectangular Components

The vectors i, j, and k are called *unit vectors* and are defined as vectors of magnitude **one** directed along the positive x, y, and z axes of a cartesian coordinate system. This is shown in Figure 2.3. A vector, A, can be resolved into rectangular components where A_x, A_y, and A_z are the scalar components along each coordinate axis as in Figure 2.4. The vector A can be defined as:

$$A = A_x i + A_y j + A_z k$$

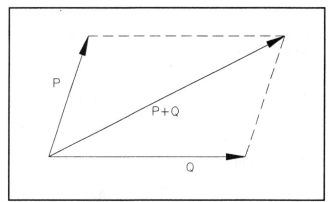

Figure 2.1 Addition of Two Vectors.

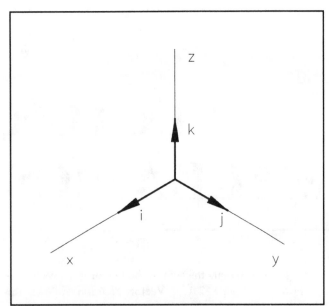

Figure 2.3 Unit Vectors.

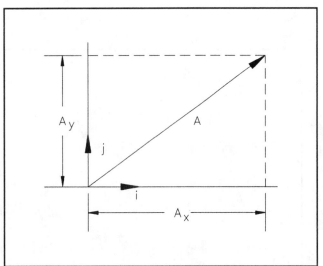

Figure 2.4 Resolution of a Vector into Rectangular Components.

Two vectors can be added by adding their rectangular components to obtain the rectangular components of the sum. Figure 2.5(a) shows two vectors A and B with their rectangular components. If the rectangular components are given by:

$$A = A_x i + A_y j$$

$$B = B_x i + B_y j$$

then the vector that is the sum of the two vectors is given by:

$$C = C_x i + C_y j = (A_x + B_x)i + (A_y + B_y)j$$

as shown in Figure 2.5(b).

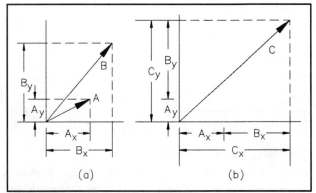

Figure 2.5 Sum of Rectangular Components of Two Vectors.

2.4 Magnitude of a Vector

A vector A can be defined as a unit vector $\hat{A}$ directed in the same orientation as A multiplied by a scalar A that indicates the *magnitude* of the vector. The magnitude of a vector can be found as:

$$|A| = \sqrt{A_x^2 + A_y^2 + A_z^2}$$

The unit vector (designated by the caret ^) is oriented in the same direction as A can be found as:

$$\hat{A} = \frac{A}{|A|}$$

Example 2.4.1. Find the unit vector that has the same direction as:

$$A = 3i + 4j$$

Solution. The magnitude of A can be found by:

$$|A| = \sqrt{3^2 + 4^2} = 5$$

Thus, the unit vector $\hat{A}$ is defined as:

$$\hat{A} = \frac{3i + 4j}{5} = 0.6i + 0.4j$$

2.5 The Product of a Scalar and a Vector

The product of a scalar k and a vector P is denoted as kP and is defined as a vector having the same orientation as P and a magnitude equal to the magnitude of P (denoted as P) multiplied by k. Multiplication of a vector and a negative scalar results in a vector having the opposite sense as illustrated in Figure 2.6.

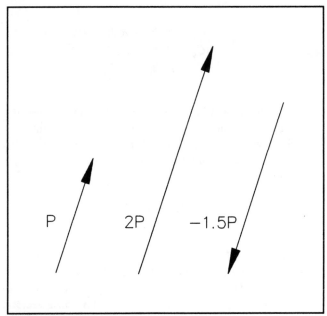

Figure 2.6 Multiplication of a Vector by a Scalar.

2.6 Scalar Product of Two Vectors

The product of two vectors is not defined in the same sense as the multiplication of scalars. One type of vector product results in a *scalar* that describes the relationship of the two vectors. The scalar product or *dot product* of two vectors A and B is defined as the product of the magnitude of the two vectors and the cosine of the angle between the two vectors, as shown in Figure 2.7. The dot product is written as:

$$A \cdot B = AB \cos\theta$$

Scalar products are both commutative and distributive.

The dot product can be readily obtained from the rectangular components of the two vectors:

$$A \cdot B = A_x B_x + A_y B_y + A_z B_z$$

Example 2.6.1. What is the angle between $A = 2i + 3j$ and $B = -1i + 4j$?

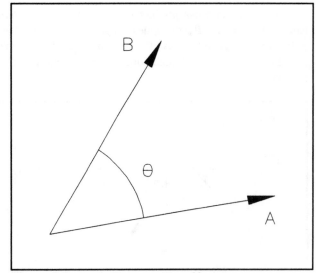

Figure 2.7 Angle Between Two Vectors.

Solution. The angle between the two vectors can be found by:

$$\theta = \cos^{-1}\left(\frac{A \cdot B}{AB}\right)$$

The dot product of A and B is found as:

$$A \cdot B = (2)(-1) + (3)(4) = 10$$

The magnitudes of the two vectors are:

$$A = \sqrt{2^2 + 3^2} = 3.61 \qquad B = \sqrt{(-1)^2 + 4^2} = 4.12$$

Therefore, the angle between the two vectors is:

$$\theta = \cos^{-1}\left(\frac{A \cdot B}{AB}\right) = \cos^{-1}\left(\frac{10}{14.87}\right) = 47.7°$$

2.7 Vector Product of Two Vectors

Another type of product that can be formed from two vectors is the vector product or *cross product*. The cross product of two vectors results in a vector that is mutually perpendicular to the two original vectors. The magnitude of the cross product is equal to the product of the magnitude of the two vectors multiplied by the sine of the angle between them:

$$|A \times B| = A \cdot B \sin\theta$$

25

The sense of the cross product is given by the *right hand rule*. If $C = A \times B$, the index finger of the right hand points in the direction of A, the second finger points in the direction of B and the thumb points in the direction of C as in Figure 2.8.

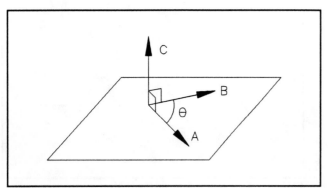

Figure 2.8 Cross Product of Two Vectors.

Cross products are distributive, but they are not commutative. Instead, if the components of the cross product are commuted the negative of the original result is obtained:

$$A \times B = -(B \times A)$$

The cross product can be evaluated by resolving A and B into their rectangular components and performing a set of algebraic operations on them. These operations can be summarized by the determinant:

$$C = A \times B = \begin{vmatrix} i & j & k \\ A_x & A_y & A_z \\ B_x & B_y & B_z \end{vmatrix}$$

The components of the cross product C can be evaluated as:

$$C_x = A_y B_z - A_z B_y$$
$$C_y = A_z B_x - A_x B_z$$
$$C_z = A_x B_y - A_y B_x$$

The cross product has many applications in engineering. For example, the moment of a force about a point O is defined as the cross product of the position vector r that gives the distance and direction of the point of application of the force F:

$$M_o = r \times F$$

Example 2.7.1. Find a vector that is perpendicular to both $A = 2i + 3j + 2k$ and $B = 1i + 4j + 3k$.
Solution. The vector that is mutually perpendicular to two vectors is given by the cross product of those vectors:

$$C = \begin{vmatrix} i & j & k \\ 2 & 3 & 2 \\ 1 & 4 & 3 \end{vmatrix} = 1i - 4j + 5k$$

Practice Problems

2.1   Determine whether the following quantities are vectors or scalars:

    a. force
    b. time
    c. velocity
    d. volume
    e. density
    f. acceleration
    g. distance
    h. angular velocity
    i. moment of a force about a point

2.2   The following vectors are defined:

$$A = 3i + 9j$$

$$B = 2i - 7j$$

$$C = -6i - 3j$$

$$D = 8i + 5j$$

Calculate the following:

    a. the sum of A and D
    b. the difference of B and C
    c. the magnitude of C
    d. the unit vector having the same direction as B
    e. the dot product of A and B
    f. the angle between $-B$ and D
    g. the cross product of B and C
    h. a vector that is mutually perpendicular to A and C

Part 2

Physics

LIGHT

Light refers to the portion of the electromagnetic spectrum that is visible to the human eye. Light can be described as an electromagnetic wave. The concept of wave propagation is used to describe the physical behavior of light.

3.1 Electromagnetic Radiation

Light is electromagnetic radiation. Light is the portion of the *electromagnetic spectrum* that is visible to the human eye. The different components of the electromagnetic spectrum are illustrated in Figure 3.1. The eye has varying sensitivity to different wavelengths of light as shown in Figure 3.2. The eye is most sensitive to yellow-green colors. This is the rationale behind the yellow color of fire trucks and self-adhesive note papers.

The speed of light in a given material is constant. The speed of light in a vacuum or air is approximately $300 \times 10^6 \, m/s$. The wave nature of light allows it to be characterized in terms of wavelength and frequency.

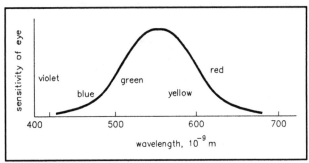

Figure 3.2 Sensitivity of the Eye to Light.

The product of the wavelength and frequency of light is equal to its speed:

$$C = \lambda \, \nu$$

where c is the speed of light in a vacuum in m/s, λ is the wavelength in m, and ν is the frequency in cycles per second or hz.

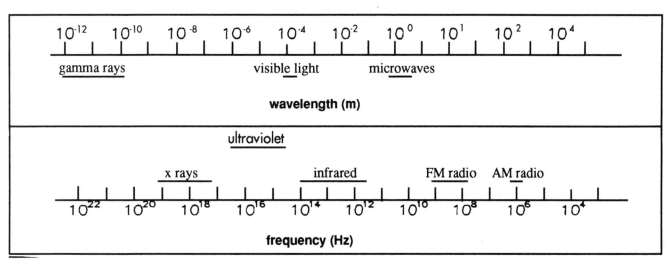

Figure 3.1 Electromagnetic Spectrum.

3.2 Ray Theory

Many aspects of the physical behavior of light can be explained in terms of *ray theory*. A ray of light is a straight path that the light travels in from one point to another. Reflection and refraction are two phenomena readily explained by ray theory.

• **Reflection.** Consider a light ray impinging on an object. A portion of the light wave is directed away from the object, while the other portions are absorbed by the object and transmitted through it. Reflected light is the portion of

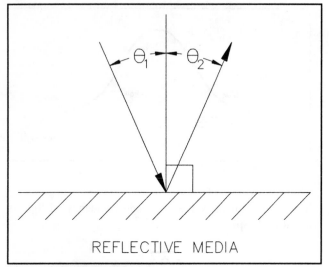

Figure 3.3 Reflected Light.

the light wave directed away from the object. If a line is drawn normal to the surface, the *angle of incidence* θ_1 (measured between the normal line and the incoming ray) is equal to the *angle of reflection* θ_2 as shown in Figure 3.3:

$$\theta_1 = \theta_2$$

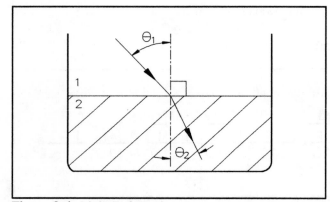

Figure 3.4 Light Refracted.

• **Refraction.** Light travels at different speeds in different media. When a ray of light is transmitted through two materials, the line of travel of the light wave is changed at the interface of the two materials. Figure 3.4 shows a ray of light travelling through one media with angle of incidence θ_1. The line of travel is changed at the interface of the two materials having an *angle of refraction* θ_2. Refraction is the basis for the action of lenses. For the situation depicted in Figure 3.4, the speed of light would be greater in material 1 than in material 2. The angles of incidence and refraction are related by *Snell's Law*:

$$\frac{\sin \theta_1}{\sin \theta_2} = \frac{v_1}{v_2}$$

where v_1 and v_2 are the speeds of light in the two media. Since it is common for the first media to be air (as in the case of lenses), it is useful to define an *index of refraction* μ:

$$\frac{\sin \theta_1}{\sin \theta_2} = \frac{c}{v_2} \doteq \mu$$

Various reflective indices are given in Table 3.1.

Table 3.1 Various Indices of Refraction

Material	μ
Water	1.33
Fused Quartz	1.46
Flint Glass	1.66
Diamond	2.42

Example 3.2.1. Light enters a pool of water from the air as in Figure 3.5. Find the angle of refraction using Snell's Law if $\theta_1 = 45°$.

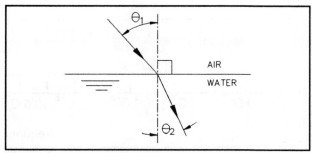

Figure 3.5 Light Refracted.

Solution. Using the data for indices of refraction, Snell's Law may be applied directly:

$$\frac{\sin \theta_1}{\sin \theta_2} = \mu$$

$$\frac{0.707}{\sin \theta_2} = 1.33, \quad \sin^{-1}\left(\frac{0.707}{1.33}\right) = \theta_2$$

$$\therefore \quad \theta_2 = 32°$$

3.3 Lenses

Lenses operate by controlling the direction of light by refraction. Most lenses can be described as a portion of two spheres joined together as in Figure 3.6. Most lenses can be considered *thin lenses*, i.e, the thickness of the lens is

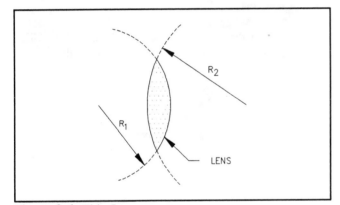

Figure 3.6 Thin Lenses.

negligible in comparison to the radius of curvature of the lens surfaces. Thin lenses can be designed and analyzed according to a few simple formulas.

The *lens equation* describes the relationship between the location of objects and images viewed by the lens, lens radii, and the index of refraction:

$$\frac{1}{p} + \frac{1}{p'} = \left(\frac{\mu_2}{\mu_1} - 1\right)\left(\frac{1}{R_1} - \frac{1}{R_2}\right)$$

where p is the distance from the lens centerline to the object viewed by the lens, p' is the corresponding distance to the image of the object created by the lens, R_1 and R_2 are the radii of the surfaces of the lens, and μ_1 and μ_2 are the indices of refraction of the surrounding media and the lens, as shown in Figure 3.7.

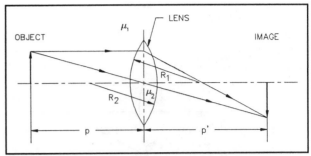

Figure 3.7 An Object Being Imaged by a Lens.

The process of selecting or designing a lens for an application typically begins with determining the *focal length*. The focal length f is the distance from the lens to the image when the object is at an effectively infinite distance from the lens (such as when the sun is being imaged). The *lens maker's equation* determines the relationship between the geometry of the lens and the focal length:

$$\frac{1}{f} = \left(\frac{\mu_1}{\mu_2}\right)\left(\frac{1}{R_1} - \frac{1}{R_2}\right)$$

The location of objects and images relative to the lens can be established by the *thin lens formula*:

$$\frac{1}{p} + \frac{1}{p'} = \frac{1}{f}$$

Important sign conventions to follow in the application of lens design equations include:

• If the center of curvature of a lens lies on the side that the ray of light is going, the radius R is positive, otherwise, R is negative.

• Distances p and f are positive when the object is on the side of the source of the light ray and the image is on the side of the destination of the ray. Otherwise both distances are negative.

Positive lens radii occur in the case of *convex* lenses. Convex lenses converge rays of light. Negative radii are applied in *concave* lenses which diverge rays of light. Figure 3.8 shows an example of each type of lens. A negative image distance occurs in the case of a concave, diverging lens such as the one show in Figure 3.8(b). Such an image is known as a *virtual image*. If an image distance is positive, a *real image* is formed.

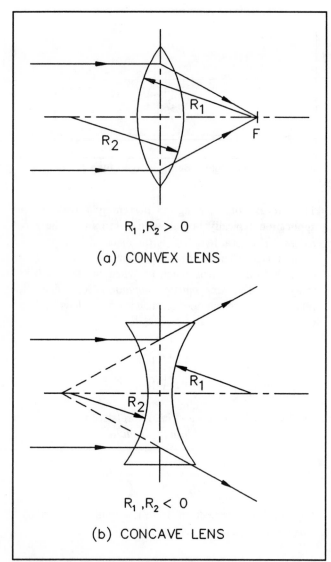

$R_1, R_2 > 0$

(a) CONVEX LENS

$R_1, R_2 < 0$

(b) CONCAVE LENS

Figure 3.8 Convex and Concave Lenses.

Example 3.3.1. A converging lens images the sun as a spot at 200 mm from the lens. Draw a ray diagram and identify the local length.

Solution. The thin lens formula can be applied with $p = \infty$:

$$\frac{1}{\infty} + \frac{1}{p'} = \frac{1}{f}$$

$$\therefore \quad f = p' = 200 \; mm$$

The associated ray diagram is shown in Figure 3.9.

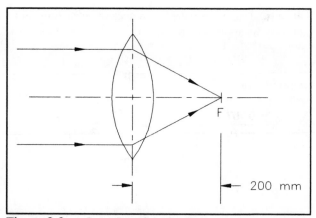

Figure 3.9 A Lens Imaging the Sun.

Example 3.3.2. The converging lens described in Example 3.3.1 is to be used to image an object placed one meter from the lens. Where will the image be formed? Is it a real image?

Solution. A ray diagram is drawn to illustrate the conditions in the problem shown in Figure 3.10. The thin lens formula is applied to find the image distance:

$$\frac{1}{1000} + \frac{1}{p'} = \frac{1}{200}$$

$$p' = 250 \; mm$$

Since the image distance is positive, a real image is formed.

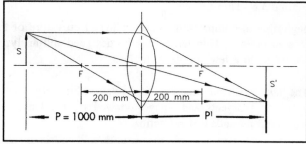

Figure 3.10 Lens Arrangement for Example 3.3.2.

Figure 3.10 shows that similar triangles can be formed based on S and S'. A formula for the magnification of a lens can be derived from these similar triangles:

$$\frac{S'}{S} = \frac{p'}{p}$$

The signs of the various quantities in the magnification equation should be ignored. Only absolute values should be used.

Example 3.3.3. A converging lens with a 100-mm focal length is to be used to form a real image of an object. The image is three times larger than the object. Where should the object be placed?

Solution. As illustrated in Figure 3.10, the image will be real. Therefore, both p and p' will be positive. The magnification equation can be used to obtain:

$$p' = 3p$$

This relationship can be substituted into the thin lens formula to obtain:

$$\frac{1}{p} + \frac{1}{3p} = \frac{1}{100}$$

Solving for p yields:

$$\frac{3p}{p(3p)} + \frac{p}{p(3p)} = \frac{3p + p}{3p^2} = \frac{1}{100}$$

$$400p = 3p^2$$

$$3p - 400 = 0$$

$$p = \frac{400}{3} = 133 \; mm$$

Practice Problems

3.1 What is the highest frequency of light that can be seen with the human eye?

3.2 What is the longest wavelength of light that can be seen with the human eye?

3.3 A manufacturer's specifications indicate that a light emitting diode is 880 nanometers (880 x 10^{-9} m). Is the emission of the LED visible? What is the frequency of the emission?

3.4 A machine shop is repainting its tools. Would blue or green be a better choice for safety? Assume a safer color is one that is more visible to the human eye.

3.5 A remote control uses an infrared light beam to turn on and off a TV set. The user wants to turn off a TV set by bouncing the beam off a mirror as shown in Figure P3.1. The remote control is positioned somewhere along the centerline of the wall. How far from the edge of the wall can the remote control be located and still turn off the set?

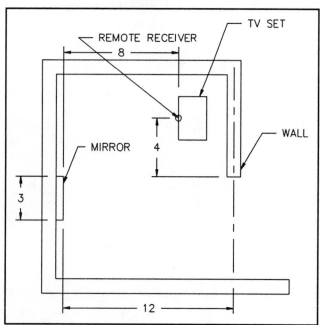

Figure P3.1 Problem 3.5.

3.6 A spotlight is located at the bottom of a decorative fountain as shown in Figure P3.2. How far from the surface of the water will the spot be projected on the wall?

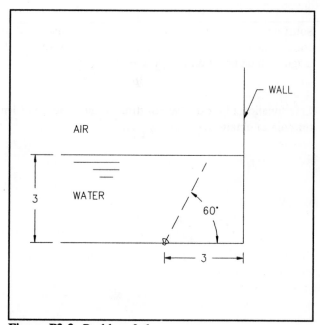

Figure P3.2 Problem 3.6.

3.7 A narrow beam of light strikes a glass plate ($\mu = 1.6$) at an angle of 53 degrees to the normal. If the plate is 4 cm thick, what will be the lateral displacement of the beam when it emerges from the plate?

3.8 An object is placed 30 cm from a converging lens with a 10 cm focal length. Find the position of the image. Is the image real or virtual?

3.9 Where must an object be placed relative to a converging lens of focal length f if the image is to be virtual and three times as large as the object?

3.10 A diverging lens has a 20 cm focal length. Where will it image an object which is 40 cm from the lens?

SOUND

Sound is the transmission of mechanical waves in matter. It can only be transmitted through matter and cannot be transmitted in a vacuum. The human ear is sensitive to certain frequencies of sound waves, typically in the range of 20 cycles/second or *hertz* (Hz) to 20,000 Hz. The ear detects mechanical vibrations in the air and the nervous system transmits them to the brain.

4.1 Wave Nature of Sound

Sound is longitudinal mechanical waves travelling through matter. Sound waves are the successive compression and rarefaction of the media that is transmitting it. The generation of sound in air can be visualized by the action of a piston in a cylinder as illustrated in Figure 4.1. If a crank is rotating at ω radians/second, a series of compressed bursts of air resulting from the forward (from left to right)

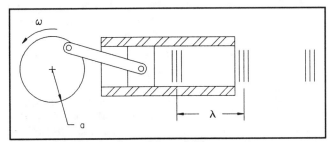

Figure 4.1 Generation of Sound Waves.

motion of the piston will be emitted from the cylinder. These compressed bursts will alternate with rarified bursts of air resulting from the retraction of the piston. The bursts will be separated by a wavelength given by:

$$\lambda = a \sin \omega t$$

where a is the radius of the crank. The frequency of the sound waves will be ω.

Sound waves propagate through a gas at a speed given by:

$$v = \sqrt{\frac{kP}{\rho}}$$

where k is the specific heat ratio for the gas, P is the pressure and ρ is the density of the gas. The values of specific heat for various gases is given in Figure 4.2.

Gas	Specific Heat Ratio (k)
Air	1.4
CO_2	1.3
CH_4	1.31
He	1.66

Figure 4.2 Specific Heat Ratio of Various Gases.

Example 4.1.1. Find the speed of sound in air under so-called "standard conditions."
Solution. Under standard conditions (room temperature at sea level), the density of air is approximately $1.29 \ kg/m^3$ and the pressure is approximately $1.01 \times 10^5 \ Pa$. Using the specific heat for air shown in Figure 4.2, the speed of sound can be found as:

$$v = \sqrt{\frac{(1.4)(1.01 \times 10^5)}{1.29}} = 331 \ m/s$$

4.2 Intensity of Sound

Sound waves represent a successive increase and decrease in pressure in the media transmitting them. This process transmits energy through the media. The *intensity* of sound P is a measure of the energy that it transmits. Intensity is defined as:

$$I = INTENSITY = \frac{ENERGY/TIME}{SURFACE\ AREA} = \frac{POWER}{AREA}$$

The intensity of sound waves passing through an open window is the power transmitted through the area of the open window.

In most cases, the intensity of sound is expressed in terms of *relative intensity or power level*:

$$RELATIVE\ INTENSITY\ (dB) = 10\log\left(\frac{I}{I_0}\right)$$

where I is the intensity of interest and I_0 is the intensity of sound at the threshold of human hearing which is typically taken as $I_0 = 10^{-12}\ W/m^2$. Relative intensity has units of *decibels* (dB).

Example 4.2.1. Find the intensity in W/m^2 of a 45 dB sound.

Solution. The intensity in decibels is the logarithmic ratio of the actual intensity to the threshold of human hearing:

$$45 = 10\log\frac{I}{I_0}$$

The actual intensity I can be found by:

$$\log\frac{I}{I_0} - 4.5$$

$$\frac{I}{I_0} = 10^{4.5} = 3.16 \times 10^4$$

$$I = 3.16 \times 10^{-8}\ W/m^2$$

The relative intensities of various types of sound are shown in Figure 4.3.

4.3 Frequency of Sound

The frequency of sound is normally referred to as its *pitch*. Pitch describes the audible effect that a frequency of sound waves has on the human ear. Pitch is normally measured in hertz (Hz) or *cycles/second*. The frequency of sound is determined by the rate of oscillation of the physical phenomena that produces the sound waves.

Sound Type	Intensity	
	W/m²	dB
Jet Aircraft (Close Range)	1	120
Jackhammer	10^{-2}	100
Automobile on Highway	10^{-4}	80
Normal Speech	10^{-6}	60
Whisper	10^{-10}	20

Figure 4.3 Typical Sound Intensities.

Example 4.3.1. A simple siren can be constructed by blowing air through a small diameter tube at a rotating disk that has a series of holes at its outer edge. If the disk has 48 holes in it and is rotating at 1200 RPM, find the pitch of the sound produced by the siren.

Solution. A single cycle of the sound wave produced by the siren occurs when a hole passes the location of the air jet. The pitch can be determined by calculating the number of times that a hole passes the blowing air in a second.

$$pitch = 1200\frac{rev}{min}\ \frac{1\ min}{60\ s}\ \frac{48\ holes\ (cycles)}{rev}$$

$$= 960\ \frac{cycles}{s} = 960\ Hz$$

4.4 Response of the Human Ear to Sound

The sensitivity of the human ear to sound is a function of frequency. The *perceived loudness* is strongly influenced by pitch. The threshold of hearing is approximately 0 dB for most sounds in the range frequencies associated with human speech (200 Hz to 5000 Hz). Frequencies lower and higher than this range must have a higher intensity before they are detected by the human ear Figure 4.4 illustrates the range of audibility of the human ear. Sounds with intensities less than the lower limit cannot be detected by an average human ear. If a sound has an intensity greater than 120 dB, it generally produces pain rather than hearing. *Ultrasonic sound* is at a frequency above the range of human hearing. *Infrasonic sound* has a frequency below the range of human hearing.

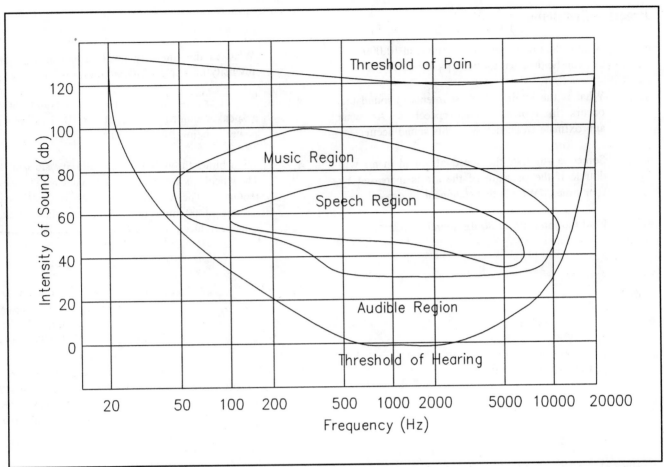

Figure 4.4 Response of Average Human Ear to Sound at Different Frequencies.

Practice Problems

4.1 What is the lowest intensity of sound at 10,000 Hz that can be detected by the typical human ear?

4.2 What is the loudest (highest intensity) sound that occurs in typical human speech. At what approximate frequency does this sound occur?

4.3 Someone suggests the speed of sound in air will double if the pressure of the air is increased by four times. Why is this statement incorrect?

4.4 Find the intensity of 80 dB sound in W/m^2.

4.5 What is the intensity level of sound with an intensity of 4×10^{-6} W/m^2.

4.6 A 2-meter by 4-meter sheet of plastic is in front of a speaker emitting sound at 90 dB. How much power is being transferred to the sheet?

4.7 Wind blows through a wheel with five slots in it. The wheel has a diameter of 18 inches with the tire mounted. The wheel is mounted on a vehicle moving at 60 miles/hour. Find the pitch of the sound produced.

THERMAL PROPERTIES OF MATTER

The thermal properties of matter are controlled by temperature. Temperature is a measure of the tendency of an object to absorb or dissipate energy in the form of heat.

5.1 Temperature Conversions

There are several scales commonly used for measuring temperature. *Temperature* measurements can be *relative* or *absolute*. Relative measurements of temperature are referenced to a physical phenomena, typically the freezing point of water. Temperature in the Fahrenheit scale T_F (part of the United States Customary System (USCS) units) is related to the Celsius scale T_C (part of the metric or Systeme International (SI) units) by the equation:

$$T_C = \frac{5}{9}(T_F - 32)$$

Absolute temperatures are referenced to the minimum achievable temperature, *absolute zero*. The value of absolute zero can be found by a simple experiment where a closed volume of gas is reduced in temperature. The pressure in the closed volume will drop with decreasing temperature. Absolute zero is the temperature found if the pressure-temperature curve is extrapolated to a pressure of zero. This temperature is a single value, independent of the gas used as shown in Figure 5.1. At this temperature (approximately $-273°C$ or $-460°F$) all atomic motion ceases, removing the kinetic energy necessary for a gas to create a pressure against a surface.

The USCS scale for absolute temperature is Rankine (T_R) which is defined as:

$$T_R = T_F + 460$$

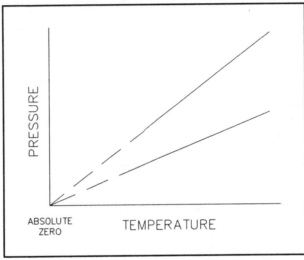

Figure 5.1 Determining the Value of Absolute Zero.

The metric scale for absolute temperature is Kelvin (T_K) which is defined as:

$$T_K = T_C + 273$$

5.2 Thermal Expansion

The dimensions of most solid materials will expand and contract with increasing and decreasing temperatures. Increasing the temperature of an object increases the motion of the atoms in the object, causing increased atomic separation and the object to grow as depicted in Figure 5.2. The change in a linear dimension, such as length or diameter, is proportional to the change in temperature of the object ΔT, its length L and a constant α, the coefficient of expansion. The change in length ΔL can be found as:

$$\Delta L = \alpha L \Delta T$$

The coefficients for some common materials are shown in Figure 5.3. The coefficients are based on a particular temperature scale. Conversion of temperatures into the appropriate scale is necessary to apply them.

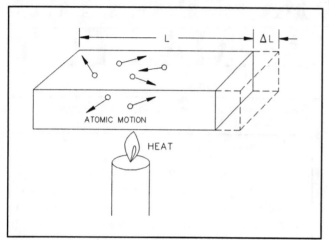

Figure 5.2 Expansion of an Object by Increase of Temperature.

Material	$\alpha \left(\frac{1}{°C} \right)$
Glass	9×10^{-6}
Concrete	10×10^{-6}
Iron	12×10^{-6}
Brass	19×10^{-6}
Aluminum	25×10^{-6}

Figure 5.3 Coefficients of Expansion.

The expansion and contraction of materials with temperature is useful for making shrink fits between parts. In addition, two strips of metal with dissimilar coefficients are bonded together to create bimetal strips widely used in thermostats. A bimetal strip will bend with a large deflection under a relatively small temperature change.

Example 5.2.1. A brass sheet has a 2.000-inch-diameter hole at 70°F (21.1°C). The sheet is heated to 300° F (149°C). Find the new diameter of the hole.
Solution. Either the coefficient of expansion or the temperatures must be converted into Celsius units to use the

values in Figure 5.3. The coefficient of expansion for fahrenheit units can be found as:

$$\alpha_F = \frac{19 \times 10^{-6}}{°C} \frac{5°C}{9°F} = \frac{10.6 \times 10^{-6}}{°F}$$

The change in diameter can be found as:

$$\Delta D = \alpha_F D \Delta T = \left(\frac{10.6 \times 10^{-6}}{°F} \right) (2.000 \ in.) \ (230° F)$$

$$= 0.005 \ in.$$

Thus, the new diameter is 2.005 inches.

5.3 Heat Capacity

The heat capacity of a material defines the amount of energy that is needed to change its temperature. Conversely, heat capacity describes the temperature change that will occur with a given amount of energy. The typical units for heat are the *calorie* (cal) and the *British thermal unit* (Btu). The units are both referenced to the heat capacity of water. One calorie is the amount of heat required to raise the temperature of one gram of water by one degree Celsius. Similarly, the Btu is the amount of heat required to raise the temperature of one pound of water by one degree Fahrenheit.

Unit heat capacity is typically quantified as *specific heat (c)* which is the quantity of heat required to change the temperature of a unit mass of substance by one degree. Specific heats are measured relative to water by definition of the units of heat. Units for specific heat are *Btu/(lb°F)* and cal/(g°C). The specific heats of various substances are shown in metric units in Figure 5.4. Using the definition of specific heat, the heat contained in a quantity of substance is given by:

$$Q = mcp \Delta T$$

where m is the mass of the substance and ΔT is the change in temperature.

Material	Specific Heat $\frac{cal}{g°C}$
Aluminum	0.21
Glass	0.20
Iron	0.11
Copper	0.09
Water	1.0

Figure 5.4 Specific Heats of Various Materials.

Specific heats are used in the study of *calorimetry*, the analysis of heat content or chemical energy in fuels, foods and other media. Known quantities of two substances, one of which is typically water, are placed in an insulated chamber known as a *calorimeter*. The heat gained by one substance is lost (or generated) by the other substance.

Example 5.3.1. How much water at $15°C$ must be used to cool a $200\ g$ part made of copper from an initial temperature of $80°C$ to a final temperature of $25°C$? Assume the contact takes place in an insulated calorimeter.
Solution. The heat lost by the copper must be the heat gained by the water. The equation relating the two quantities of heat is:

$$(mc\Delta T)_{copper} = (mc\Delta T)_{water}$$

The unknown mass of water can be found by:

$$(200g)(0.093cal/g°C)(80-25°C) = m_{water}(1)(25-15)$$

$$m_{water} = 102\ g$$

5.4 Thermodynamics

Thermodynamics is the study of energy in transition. A knowledge of thermodynamics is critical in analyzing the operation of steam power plants, refrigerators and other devices associated with energy. An extensive study of thermodynamics is beyond the scope of this book. However, some basic ideas in thermodynamics will be described to provide a more complete discussion of thermal sciences.

There are two basic physical laws of thermodynamics which can be applied to all processes involving heat, work and energy. The *First Law of Thermodynamics* is commonly known as conservation of energy. Energy cannot be created nor destroyed, it can only be changed in form. In the context of thermodynamics, this law is stated as:

$$Q = \Delta U + W$$

where Q is a quantity of heat, ΔU is a change in internal energy and W is the work performed.

The consequences of the first law can be illustrated by a quantity of gas contained in a cylinder with a perfectly sealing piston as shown in Figure 5.5. In Figure 5.5(a), work is done on the system as the piston is being forced down. This work input can result in an increase in the internal

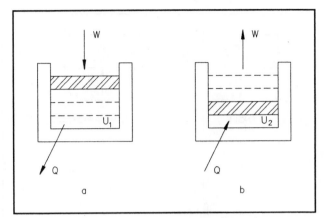

Figure 5.5 The First Law of Thermodynamics.

energy of the gas (U) (as evidenced by an increase in temperature) and heat (Q) being transferred out of the cylinder. If the cylinder is insulated and no heat transfer occurs, all of the work input will result in raising the internal energy of the gas, causing a rise in its temperature and pressure. A smaller change in internal energy will occur if heat exchange to the surroundings occurs. In Figure 5.5(b), heat is added to the system from the surroundings. The heat will cause the gas to expand, doing work by raising the piston. If the piston is prevented from being raised, all of the heat input will result in a change in internal energy of the gas, causing a rise in temperature and pressure.

The *Second Law of Thermodynamics* describes the relationship of work and heat. One consequence of the second law is that heat flows spontaneously from a hot object to a cold object and not vice versa. Work must be done to transfer heat from a cold object to a hot object. One quantity of heat Q_H may be extracted from a hot object at temperature T_H and a lessor quantity of heat Q_L will be dissipated into a corresponding cold object at T_L. The difference between the two quantities of heat can be captured as useful work W_{OUT} as illustrated in Figure 5.6(a).

$$W_{OUT} = Q_H - Q_L$$

An engine operates under these conditions, moving heat from a fuel, performing work and dissipating heat to the surroundings.

A quantity of heat Q_L can be extracted from a cold object and through the action of the input of work, a larger quantity of heat Q_H is dissipated to the surroundings.

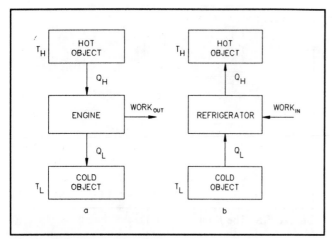

Figure 5.6 Thermodynamic Cycles.

The quantity of work required to move the heat is given by:

$$W_{IN} = Q_H - Q_L$$

A refrigerator operates under these conditions, requiring work input to move heat from the interior of the refrigerator to the surroundings as shown in Figure 5.6(b).

5.5 Heat Transfer

Heat can be transferred between two objects in three modes, *conduction, convection and radiation.* Heat will spontaneously flow from a hot object to a cold object through one or more of these three modes.

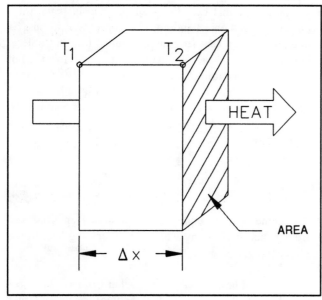

Figure 5.7 Heat Conduction.

Heat conduction describes the transfer of energy from a high-temperature region to a low-temperature region through a solid object, as shown in Figure 5.7. The rate of the energy transfer is dependent on the area of conduction **A**, temperature difference between the hot and cold regions, ($\Delta T = T_1 - T_2$) and the thickness of the material Δx. The rate of transfer, given in Btu/s, cal/s, or watts, is given by:

$$\frac{\Delta Q}{\Delta t} = kA \frac{\Delta T}{\Delta x}$$

where k is the thermal conductivity or the conduction coefficient. Typical values of k are given in Figure 5.8.

Material	$k \dfrac{cal}{s^\circ C\, cm^2}$
Silver	1.00
Copper	0.92
Aluminum	0.50
Glass	20×10^{-4}
Rubber	5.0×10^{-4}

Figure 5.8 Specific Heats of Various Materials.

Example 5.5.1. A copper rod is 10 cm long and 4 cm in diameter. One end of the rod is at $1000^\circ\, C$ and the other end of the rod is at $70^\circ\, C$. Find the rate of heat transfer from one end of the rod to the other.

Solution. Using the conduction coefficient for copper found in Figure 5.8, the conduction equation can be applied directly:

$$\frac{\Delta Q}{\Delta t} = 0.92 \left(\frac{\pi 4^2}{4} \right) \frac{930}{10} = 1075 \frac{cal}{s}$$

Heat convection describes the transfer of energy from a surface by the flow of a fluid over an object. Figure 5.9 shows heat loss from a wall by the flow of a fluid. The fluid velocity has a *gradient* such that it has zero velocity at the wall and reaches maximum velocity v_∞ in the *freestream* where the flow of the fluid is unaffected by the presence of the wall. The temperature of the wall is T_w and the temperature of the freestream is T_∞. Under these conditions, Newton's law of convection heat transfer is given by:

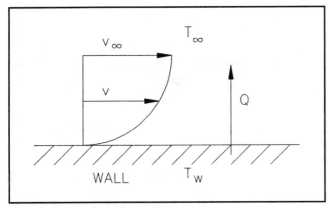

Figure 5.9 Heat Transfer by Convection.

$$\frac{\Delta Q}{\Delta t} = hA(T_W - T_\infty)$$

where h is the convection coefficient. The convection coefficient varies widely depending on the physical situation. Typical values for the coefficient are given in Figure 5.10.

Mode	$h \quad \dfrac{W}{m^2 \cdot C}$
Vert. Plate in Air	4.5
Horiz. Cylinder in Air	6.5
Forced Air Over Plate	75
Boiling Water	5000

Figure 5.10 Convection Coefficients for Various Situations.

Example 5.5.2. An electrical resistance heater is used to boil water. The diameter of the wire is 1 mm and the length of the wire is 10 cm. The convection coefficient is $5000 \ W/(m^2 \, {}^\circ C)$. Find the electrical power that must be supplied to keep the surface of the wire at $114\,^\circ C$.
Solution. The heat transfer occurs between a wire surface that is at $114\,^\circ C$ and a freestream that is at the boiling temperature of water $100\,^\circ C$. The convection area is the surface area of the wire:

$$A = \pi dL = \pi (1 \times 10^{-3} m)(10 \times 10^{-2} m) = 3.14 \times 10^{-4} \ m^2$$

The rate of energy transfer or power can be found as:

$$Q = (5000)(3.14 \times 10^{-4})(114 - 100) = 22 \ W$$

The third mode of heat transfer is radiation. Electromagnetic radiation carries energy from one body to another. The rate of energy transfer is based on the performance of ideal radiators and absorbers called *black bodies*. An ideal black body would be an infinitely large black plate. Other objects with different geometries and emissivities (the ability to emit radiation) determine the amount of heat transfer that will occur.

The rate of energy transfer in radiation between two objects is given by:

$$q = k \sigma A(T_1^4 - T_2^4)$$

where k is the radiation constant ($k = 1$ for two infinite black plates) which depends on geometry and emissivity, A is the area of emission and absorption, T_1 and T_2 are the absolute temperatures of the two objects and σ is the Stefan-Boltzman constant. The Stefan-Boltzman constant is:

$$5.6 \times 10^{-8} \ W/(m^2 \, {}^\circ K^4)$$

or

$$0.17 \times 10^{-8} \ Btu/(hr \ ft^2 \, {}^\circ R^4)$$

Example 5.5.3. Two large black plates are at $800\,^\circ C$ and $300\,^\circ C$. Find the heat transfer per unit of area.
Solution. The two plates will be assumed to act as ideal black bodies, therefore $k = 1$. The radiation heat transfer equation can be applied as:

$$\frac{\Delta Q/\Delta t}{A} = k\sigma (T_1^4 - T_2^4)$$

If the known temperatures in degrees Kelvin and the Stefan-Boltzman constant are substituted, the following result is obtained:

$$\frac{\Delta Q/\Delta t}{A} = 1.0)(5.6 \times 10^{-8})(1073^4 - 573^4) = 68 \ kW/m^2$$

Practice Problems

5.1   Convert a temperature of 295° F into the following temperature scales:
     a.   Celsius
     b.   Rankine
     c.   Kelvin

5.2   A shaft is made of iron. It has a diameter of 1.5000 inches at 70° F. Find the diameter at 212° F.

5.3   A shrink fit is needed for two mating aluminum parts (a shaft and a hole). The diameter of the hole is 2.000 inches and the shaft is 2.002 inches at 70° F. Only the female part will be heated. What temperature should the female part be heated to for 0.001 inches of clearance at assembly?

5.4   An iron part is being heat treated. The part will be heated to 1200° F and plunged into water at 55° F. Two kilograms of water will be used. The part has a mass of 250 g. What is the final temperature of the water and the part?

5.5   A water-to-water heat exchanger is used to cool water used in a welding operation. The water used for welding is at 150° F and is flowing at 5 kg per minute. It must be cooled to 90° F by water at 60° F. What mass flowrate of the low temperature water is necessary to achieve the desired cooling if the heat exchanger is 100% efficient?

5.6   A quantity of gas is contained in a cylinder with a perfectly sealing piston. The piston performs five units of work by compressing the gas. A total of four units of heat were transferred to the surroundings of the cylinder. Did the internal energy of the gas increase or decrease?

5.7   A refrigerator manufacturer claims that running the refrigerator in your house will not increase the load on your air conditioning. What is wrong with this statement?

5.8   Which is the better thermal conductor, aluminum or copper?

5.9   A glass window in a house is 1 meter by 1 meter in area and 5 mm thick. The outside temperature of the glass is -5°C, while the internal temperature of the house is 20°C. What is the heat loss from the house, per second, through the window?

5.10   A copper tube with a diameter of 10 mm and a wall thickness of 1 mm carries pressurized water at 200° C. A 50 mm length of the tube passes through a container filled with 0.5 kg of water initially at 15° C. How long will it take for the water to reach 30° C? Neglect convection effects.

5.11   A pan of water is 100 mm in diameter. The water is 50 mm deep in the pan. When the water is placed on a stove and the water is boiling, the surface of the pan in contact with the water is at 150° C. Find the rate of energy transfer to the water.

ELECTRICITY AND MAGNETISM

Electricity and magnetism are interrelated phenomena. They are involved in the generation, transmission and storage of power in numerous applications. The field of study in these topics is very large. The coverage here will be limited to a brief discussion of circuits, power flow, electrostatics and electromagnetism.

6.1 Circuits

Electrical circuits are the interconnection of components for generating and distributing electrical power, converting electrical power to another form (such as light, heat, or motion), or processing information. Electrical circuits contain a source of electrical power, passive components which dissipate or store energy, and active components which change the form of electrical power. Circuits can be broadly classified as *direct current* (DC) where currents and voltages do not vary with time and as *alternating current* (AC) where currents and voltages vary (usually sinusoidally) with respect to time.

There are several quantities that are used in electrical circuits:

• Charge (Q). Electrical charge is an energy carrying quantity that is measured in units of *coulombs*. The smallest unit of electrical charge is the *electron* which carries a charge of 1.6022×10^{-19} *coulombs*. Charges are quantized into plus and minus units. Opposite charges are attracted to each other. Like charges repel each other. Charges exert a force on each other. This force is the basis of electrical power.

• Current (I). Electrical current is the time rate of flow of charge past a point in a circuit, Current is measured in *Amperes*. In an analogy to a fluid system, electrical current is analogous to the volumetric flowrate of water through a

pipe. A higher electric current implies a greater "volume" of energy being delivered in a given period of time.

• Voltage (V). Voltage is the change in energy per unit charge. The unit of measure is the *volt*. Voltage can increase or decrease as current flows through circuit elements. Batteries and other sources can increase voltage. Resistors and other loads decrease voltage as current flows through them. Voltage is analogous to pressure in a fluid system.

• Energy (W). Electrical energy is the capacity to do work. Energy is measured in *joules*. Electrical energy can be stored in circuit elements such as batteries, capacitors or coils. Electrical energy can be transformed into mechanical energy in a motor or dissipated as heat through a resistor.

• Power (P). Electric power is the time rate of energy flow. Electrical power is measured in *watts*. The energy consumed by a household in a billing cycle is commonly expressed in units of kilowatt·hours, indicating an amount of power for a period of time. This unit is actually a number of joules per second. The power supplied by a circuit component can be found as the product of the voltage rise across the component and the current that flows through it:

$$P = IV$$

The power consumed by a component can be found as the product of the voltage drop across the component and the current that flows through it:

$$P = IV$$

Alternately, the power consumed in a resistive load can be found as the product of the resistance and the square of the current that flows through it:

$$P = I^2 R$$

There are many types of components that can be used to form an electrical circuit. Some examples of sources that generate electrical energy are: batteries, generators and power supplies (devices that convert one type of voltage/current combination into another). Some of the passive components that are used in circuits are: resistors, inductors and capacitors. These components are illustrated in Figure 6.1 in the context of a DC circuit.

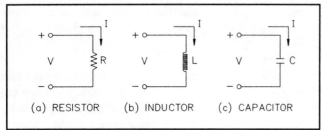

Figure 6.1 Circuit Components.

• *Resistors* are energy absorbing components. A resistor is symbolized in Figure 6.1(a). Resistance is measured in *ohms* (Ω). The relationship between current and voltage in a resistor is given by:

$$I = \frac{V}{R} \quad or \quad AMPERES = \frac{VOLTS}{OHMS}$$

• *Inductors* are energy storing components where energy is stored in a magnetic field. An inductor is illustrated in Figure 6.1(b). Inductance is measured in *henries* (L). The relationship between current and voltage in an inductor is given by:

$$I = \frac{1}{L}\int Vdt \quad or \quad AMPERES = \frac{VOLT \cdot SECONDS}{HENRIES}$$

• *Capacitors* are energy storing components where energy is stored in an electric field. A capacitor is shown in Figure 6.1(c). Capacitance is measured in *farads* (F). One farad is very large. A typical capacitor used in a circuit for consumer electronics has a value measured in the *microfarads* (μF). The relationship between current and voltage in a capacitor is given by:

$$I = C\frac{dV}{dt} \quad or \quad AMPERES = \frac{FARAD \cdot VOLT}{SECONDS}$$

6.2 Types of Circuit Connections

The two basic types of circuit connections, parallel and series, are illustrated in Figure 6.2. In a *parallel connection*, the same voltage is present across all components. In a *series connection*, the same current flows through all components. Components in a circuit may be combined and analyzed as a simpler circuit containing fewer components. The rules for combining circuit components vary depending on the type of component and whether they are connected in parallel or series. An equivalent component can be found that has the same performance as a set of components in the circuit. The rules for combining components are summarized in Figure 6.3. Some comment should be made about combining sources in parallel and series. Sources can be combined in series to obtain an additive equivalent voltage. Sources having equal voltages can be combined in parallel to supply a large current without demanding an excessive current from any single source. The equivalent voltage in a parallel connection of equal voltage sources is equal to that of any individual source. Sources having appreciably different voltages are not connected in parallel since wasteful circulating currents would occur, even if there was not external connection to the sources.

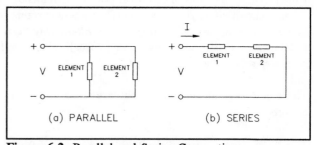

Figure 6.2 Parallel and Series Connections.

Example 6.2.1. Find the single equivalent resistance to the interconnected resistors shown in Figure 6.4.
Solution. The circuit can be reduced to an equivalent resistance by repeatedly applying the rules for combining resistances in series and in parallel. The $6\,\Omega$ and $3\,\Omega$ resistors are combined in parallel. The $4\,\Omega$ and the resulting $2\,\Omega$ resistors are combined in series. The resulting parallel branch of $6\,\Omega$ and $12\,\Omega$ is combined in parallel. The resulting $4\,\Omega$ and $5\,\Omega$ are combined in series to obtain an overall equivalent resistance of $9\,\Omega$.

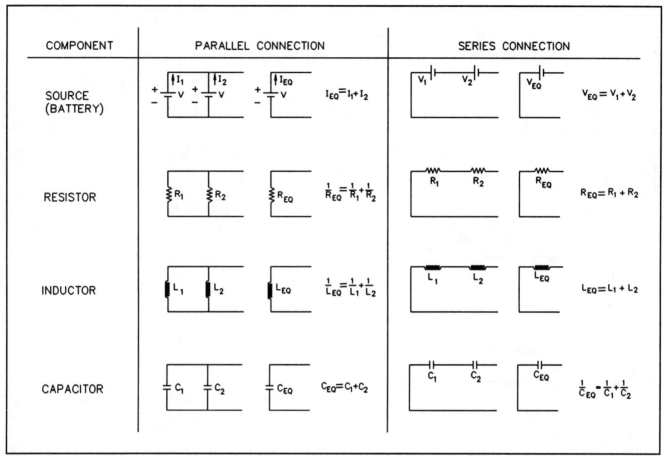

Figure 6.3 Parallel and Series Connections of Various Components.

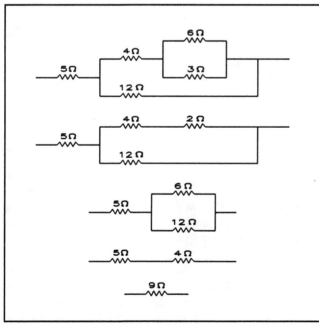

Figure 6.4 Network of Resistors.

6.3 Circuit Analysis Using Kirchoff's Laws

There are two tools used in analyzing simple electrical circuits. These tools are based on two conservative laws that govern the behavior of basic circuits:

• *Kirchoff's Loop Rule* (KLR) is a statement of conservation of energy. It states that the sum of voltage rises or drops around a closed path or loop must be zero.

• *Kirchoff's Point Rule* (KPR) is a statement of conservation of charge. It states that the flow of charges (current) into or out of a point (junction of electrical connections) must add to zero.

Example 6.3.1. For the circuit shown in Figure 6.5, find the current I, the power sourced by the batteries, and the power dissipated by each resistor.

Solution. Kirchoff's loop rule is applied with a current I flowing counterclockwise around the path. The sum of the voltage changes are set equal to zero to obtain:

49

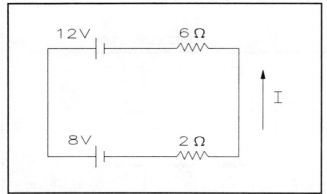

Figure 6.5 Circuit for Example 6.3.1.

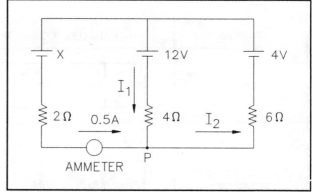

Figure 6.6 Circuit for Example 6.3.2.

$$\Sigma V_{CHANGES} = 0: \quad 12 - 8 - 2I - 6I = 0$$

$$I = 0.5 \, A$$

It should be noted that the voltage change across the eight-volt source is taken as negative since the current flows into the positive side of the source and exits through the negative side. The power sourced by the batteries is the sum of the product of current and voltage through each:

$$P_{SOURCED} = \Sigma IV = (0.5)12 - (0.5)8 = 2 \, W$$

The eight-volt source is storing energy, possibly being charged if it were a rechargeable battery. The power dissipated by the circuit is given by:

$$P_{DISSIPATED} = \Sigma I^2 R = (0.5)^2 6 + (0.5)^2 2 = 2 \, W$$

Clearly, all of the power being supplied by the source is being dissipated by the resistors.

Example 6.3.2. For the circuit shown in Figure 6.6, find the voltage that must be supplied by source X if the ammeter indicates that a 0.5-ampere current is flowing in the circuit.
Solution. Kirchoff's Point Rule is applied at P. The 0.5-ampere current and I_1 flow into the junction and I_2 flows out of the junction.

$$0.5 + I_1 - I_2 = 0$$

Kirchoff's Loop Rule is applied in the left loop:

$$-X - (0.5)(2) + 4I_1 + 12 = 0$$

Similarly, the Loop Rule is applied in the right loop:

$$-12 - 4I_1 - 6I_2 + 4 = 0$$

Solving the three equations simultaneously yields:

$$I_1 = -1.1 \, A$$

$$I_2 = -0.6 \, A$$

$$X = 6.6 \, V$$

Note, the negative signs on the currents indicate that the directions shown in Figure 6.6 are opposite to the actual flow of current in each case.

6.4 Electric Fields and Electricity

Two electric charges or objects carrying an electric charge exert forces on each other. This force is proportional to the inverse square of the distance between the charges. Charge is measured in Coulombs (C). The smallest amount of charge is that carried by an individual electron which is $1.6022 \times 10^{-19} \, C$. This force relationship is known as Coulombs Law and is given by:

$$F = k\frac{q_1 q_2}{r^2}$$

where q_1 and q_2 are the amount of charge that is carried by the two objects, k is the permittivity constant with a value of $9.0 \times 10^9 \, N \cdot m^2/C$, r is the distance between the objects in meters, and F is the force acting between the objects in Newtons.

Example 6.4.1. Two small pieces of cellophane wrapping are electrically charged positive as they are torn off a package. The pieces are separated by $0.6 \, m$. What is the force acting between them if each piece carries $0.24 \, \mu C$ of charge?

Solution. Coulomb's law can be applied directly:

$$F = (9.0 \times 10^9)\frac{(0.24 \times 10^{-6})(0.24 \times 10^{-6})}{0.6^2}$$

$$= 1.4 \times 10^{-3} \, N$$

Electric charges behave differently depending on the material that carries the charge. *Conductors* are materials in which electric charges are free to move. *Insulators* are materials in which electric charges are fixed. *Electric current*, measured in amperes (A), is a measure of the amount of charge that flows through a conductor in a unit time.

Electric potential or *voltage* is the ability to do work by moving a charge and is measured in J/C or volts (V). *Potential difference* describes the strength of an electric field over a distance. Potential difference is measured as a difference in voltage between two conductors and is given by:

$$V_B - V_A = Ed$$

where V_A and V_B are the voltages that are present at the two conductors, d is the distance between the conductors in meters and E is the electric field strength in V/m.

Example 6.4.2. Find the electric field strength between the parallel plates shown in Figure 6.7.

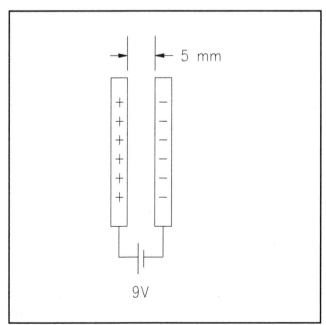

Figure 6.7 Parallel Charged Plates.

Solution. The potential difference between the plates is nine volts. The electric field strength can be found by:

$$V_B - V_A = Ed$$

$$9V = E(0.005 \, m)$$

$$E = 1800 \, V/m$$

6.5 Magnetic Fields

Two magnetic objects exert force on each other. A field exists around a magnetic object that defines how the force will be exerted. The field is made of *lines of flux* that represent the direction that a compass needle would be oriented in if it were placed in the field. The source of the magnetic field is a moving charge. In the case of a permanent magnet, the spinning electrons in the atom provide the moving charge. There are domains of material that are similarly oriented, providing a net magnetic field. In the case of an electromagnet, a wire carrying current (a moving charge) generates the magnetic field as shown in Figure 6.8.

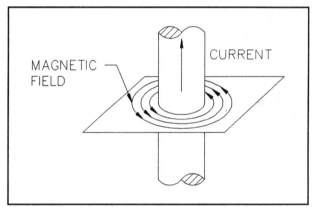

Figure 6.8 The Magnetic Field Resulting From Current Flowing Through a Conductor.

The magnetic field strength determines the amount of force that will be exerted on an object by a magnetic field. Magnetic field strength B is measured in units of teslas or gauss. One tesla is equivalent to 10^4 gauss where one gauss is the approximate magnetic field of the earth. A tesla is a derived unit given by:

$$1 \, tesla = \frac{Newtons}{Amps \cdot Meters}$$

The force F exerted on a wire carrying current I is illustrated in Figure 6.9. The force exerted on the wire is given by:

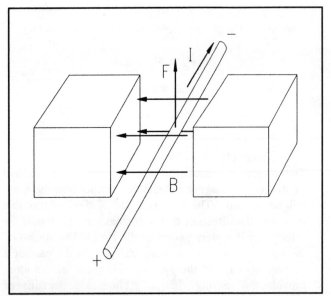

Figure 6.9 Force Exerted on a Current-Carrying Conductor in an Magnetic Field.

$$F = BIL$$

where F is the force in newtons, B in the magnetic field in teslas, I is the current in amperes and L is the length of the wire in the field in meters. The direction of the force is given by the right hand rule. If your index finger points in the direction of the current and your middle finger points in the direction of the magnetic field, the force exerted on the wire acts in the direction of your thumb.

Example 6.5.1. A 1-meter-long wire runs east-west and carries a current of 20 amperes in the east-to-west direction. Find the magnitude and direction of the force exerted on the wire.

Solution. Since the current runs from east to west and the magnetic field points north, the direction of the force will be oriented toward the earth as given by right hand rule. The magnitude of the force is given by:

$$F = BIL = \left(10^{-4}\frac{N}{A\cdot m}\right)(20\,A)(1\,m) - 0.002\,N$$

Practice Problems

6.1 An automotive electrical system operating at 12 volts contains a light bulb that draws 2 amperes. What is the power consumed by the light bulb?

6.2 A resistor rated at 100 ohms is placed across the terminals of a 12-volt battery. How much current does the resistor draw from the battery?

6.3 An household electric heater operates at 110 volts. The unit is rated at 1000 watts. What is the resistance of the heater?

6.4 What physical quantity is measured in kilowatt-hours?

6.5 Two 1.5-volt batteries are to be connected to supply a total of 3.0 volts to an electrical load. Should the batteries be connected in series or parallel?

6.6 A group of five capacitors is rated at $5\mu F$ each. If they are all connected in series, what will their equivalent capacitance be?

6.7 Find the total equivalent resistance between points A and B in the circuit shown in Figure P6.1.

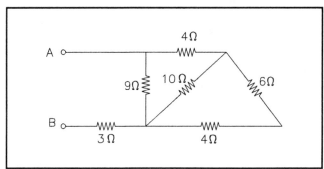

Figure P6.1 Problem 6.7.

6.8 Find the current I in the circuit shown in Figure P6.2.

6.9 Find the current I in the circuit shown in Figure P6.3.

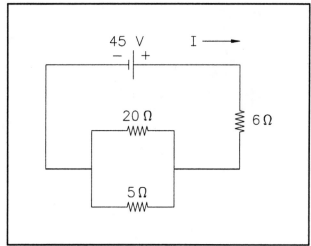

Figure P6.2 Problem 6.8.

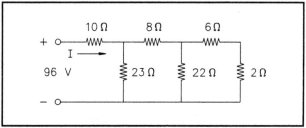

Figure P6.3 Problem 6.9.

6.10 A 12-volt battery is attached to two parallel plates. The positive terminal is connected to one plate, the negative to the other. The plates are separated by 1.5 mm. Find the electric field between the plates.

6.11 The rear window defroster in a car will create a slight magnetic field when it is turned on. A car is travelling north. The rear window can be considered vertical. The current flows from the passenger side to the drivers side of the rear window. Determine the direction of the magnetic force exerted on the defroster wires.

Part 3

Engineering Sciences

STATICS

Statics is the analysis of the mechanical equilibrium of rigid bodies subjected to force systems. The term *statics* is used because the analysis is restricted to bodies at rest. Traditionally, engineering statics requires an understanding of forces, the transmissibility of forces, Newton's Laws, and the construction of free body diagrams. Centroids, centers of gravity, and moments of inertia are also typically included in a discussion of statics.

7.1 Force

Force is a vector quantity. It is specified by both a magnitude and a direction. The study of statics depends on several basic principles:

- **Transmissibility**

Transmissibility is the principle that the equilibrium of a rigid body will remain unchanged if a force F on a rigid body is replaced by a force F' with the same magnitude, direction and same line of action acting at a different point. A force may be transmitted along its line of action without changing the effect that it has on a body as seen in Figure 7.1.

- **Parallelogram Law**

Two forces acting on a particle can be replaced by a single force known as the *resultant*. The resultant is obtained by vector addition of the two forces. This can be visualized

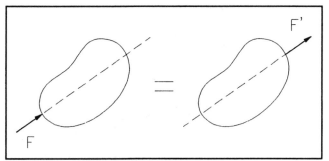

Figure 7.1 Transmissibility of Forces.

as drawing the diagonal of the parallelogram having sides given by the force vectors shown in Figure 7.2.

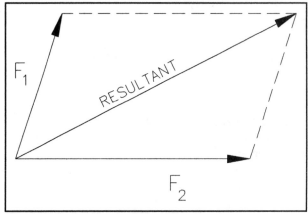

Figure 7.2 Resultant of Two Forces.

- **Newton's Laws**

First Law. A particle will remain at rest or will move in a straight line at a constant speed when the resultant force acting on the particle is zero. This can be summarized as:

$$\Sigma F = 0$$

This equation can be read as "the summation of all forces acting on a particle is equal to zero."

Second Law. If the resultant force acting on a particle is not equal to zero, the particle will have an acceleration with a magnitude that is proportional to the resultant force and a direction along the resultant force. The constant of proportionality is the *mass* of the particle, m. This law is summarized as:

$$F = ma$$

This concept is described in greater detail in Section 8, Dynamics.

Third Law. The forces of action and reaction between bodies in contact have the same magnitude, same line of action and opposite sense. Figure 7.3 shows a block being pushed against a wall with force F applied on the left side. The wall reacts against the right side of the block with an equal force, with opposite sense along the same line of action.

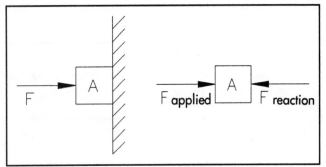

Figure 7.3 Reaction to an Applied Force.

7.2 Rectangular Components of a Force

It is often useful to resolve a force into components that are along perpendicular coordinate axes. In Figure 7.4, the force F has been resolved into two rectangular components. The magnitudes of these components are given by:

$$F_x = F \cos \theta$$

$$F_y = F \sin \theta$$

This result is often useful in adding together the forces acting on a particle. This is the analytical equivalent to

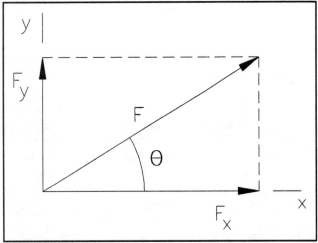

Figure 7.4 Rectangular Components of a Force.

the parallelogram law. The rectangular components of the resultant force acting on a particle are given by:

$$R_x = \Sigma F_x$$

$$R_y = \Sigma F_y$$

Example 7.2.1. Three forces act on the eyebolt as shown in Figure 7.5. Find the resultant force.

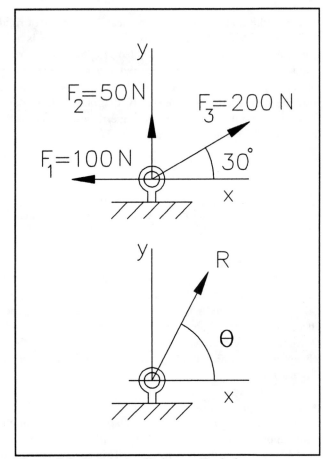

Figure 7.5 Forces Applied to an Eyebolt.

Solution. Refer to Table 7.1.

$$\tan \theta = \frac{R_y}{R_x} = \frac{150}{73}$$

$$\theta = 64°$$

$$R = \sqrt{73^2 + 150^2} = 167 \, N$$

or

$$R = 167 \, N \angle 64°$$

58

Table 7.1 Forces Applied to an Eyebolt.

Force	Magnitude N	x Component N	y Component N
F_1	100	-100	0
F_2	50	0	50
F_3	200	$200 \cos 30° = 173$	$200 \sin 30° = 100$
		$R_x = 73$	$R_y = 150$

7.3 Moment of Force

A *moment* is the tendency to rotate that a force imparts to a rigid body. A moment about a point is defined as the cross product of a distance from the point of interest to the line of action of a force. In Figure 7.6 the moment about point O is given by the cross product:

$$M_O = r \times F$$

The vector-valued moment, M_O, is directed out of the page. The magnitude of the moment is given by $r \sin \theta \, F$. The distance term, $R \sin \theta$, is the perpendicular distance from point O to the line of action of the force, d.

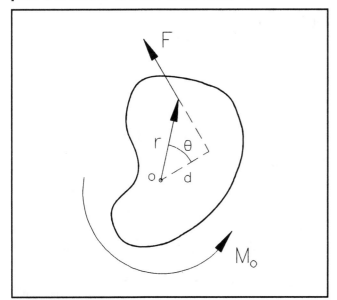

Figure 7.6 Moment of a Force About a Point.

7.4 Force Couples

Two forces of equal magnitude and opposite sense with parallel lines of action form a *couple*. Figure 7.7 shows two forces acting on an object. These two forces can be resolved into a moment of magnitude M_O given as:

$$M_O = (d)(F)$$

where d is the perpendicular distance between the lines of action of the two forces. A couple will result when there is a moment generated by two balanced forces.

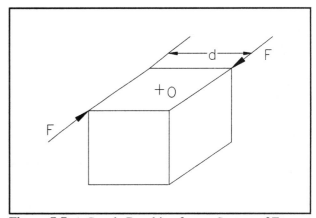

Figure 7.7 A Couple Resulting from a System of Forces.

7.5 Forces and Couples

A force acting on a body can be considered equivalent to a force acting on a specified point and a couple about that point. This is often a useful tool in analyzing statics

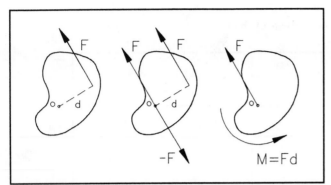

Figure 7.8 An Equivalent System of Forces and Moments.

problems. Figure 7.8 shows a force, F, acting at a distance of d from point O. Two forces of a magnitude equal to F with opposite sense can be superimposed at point O with no net change on the static equilibrium of the body. This system of forces can be resolved into a force acting through point O and a moment acting about point O.

7.6 Newton's First Law and Moments·

Newton's First Law may be extended to a rigid body with one additional observation about static equilibrium. As in the case of a particle, the summation of all forces acting on the body must sum to zero. In addition, the summation of all moments acting about any point in the body must sum to zero. These statements may be expressed as:

$$\Sigma F = 0$$

$$\Sigma M = 0$$

7.7 Free Body Diagrams

A *Free Body Diagram* (FBD) is the technique used to identify the relevant forces and moments that affect a body. When drawing an FBD, the body is isolated from all contacting bodies. Next, all forces acting on the body, and external reactions are depicted. The general procedure for creating an FBD is:

• Indicate all external forces acting on a body,

• Isolate the body from the ground or any bodies in contact with it and

• Identify the magnitude and direction of reactions from the ground or other bodies in contact by the application of Newton's First Law.

In all cases, a set of appropriate coordinate axes that are fixed to the body should be selected.

Example 7.7.1. A simple structure that is fixed to the ground is shown in Figure 7.9. Find the reactions at the ground.

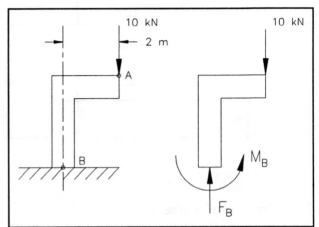

Figure 7.9 Loaded Structure and Corresponding Free Body Diagram.

Solution. An FBD is drawn showing a force F_B reacting against the structure at B. A moment M_B also reacts against the structure that prevents it from rotating. If the forces are summed in the vertical direction and set equal to zero (a sign convention will be used where forces acting upward will be taken as positive)

$$\Sigma F_V = 0 = F_B - 10kN$$

then: F_B is found to be $10\,kN\,\uparrow$. Similarly, if the moments acting on the structure are summed at point B and set equal to zero (using a sign convention where counterclockwise moments are taken as positive), then the 10 kN load acting at 2 m from B has a reaction at the ground given by:

$$\Sigma M_B = 0 = M_B - (10kN)(2m)$$

M_B is found to be $20\,kN \cdot m$ (counterclockwise).

7.8 Centers of Gravity and Centroids

A force that is distributed on a body can be treated as an equivalent force acting at a single point. The point of application of the resultant force is known as the *centroid*. If the distributed force is the weight of the object, the centroid is referred to as the *center of gravity*. The center of gravity is the location where all the mass in an object

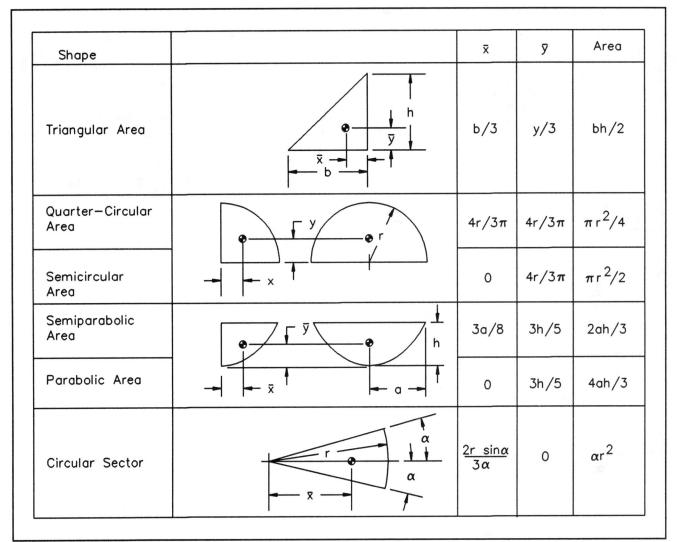

Shape		$\bar{x}$	$\bar{y}$	Area
Triangular Area		b/3	y/3	bh/2
Quarter–Circular Area		$4r/3\pi$	$4r/3\pi$	$\pi r^2/4$
Semicircular Area		0	$4r/3\pi$	$\pi r^2/2$
Semiparabolic Area		3a/8	3h/5	2ah/3
Parabolic Area		0	3h/5	4ah/3
Circular Sector		$\dfrac{2r\ \sin\alpha}{3\alpha}$	0	αr^2

Figure 7.10 Centers of Gravity.

can be considered to be concentrated. Figure 7.10 shows the location of the center of gravity of some common two-dimensional objects. The center of gravity may be found by subdividing the object into subcomponents where the center of gravity is known and identifying the location of the centroid of each piece relative to a datum (reference line) or reference point. In a two-dimensional problem, the location of the center of gravity in the y direction, $\bar{Y}$, relative to the datum can be found by:

$$\bar{Y} = \frac{\sum\limits_{i-1}^{N} \bar{y}_i A_i}{\sum\limits_{i-1}^{N} A_i}$$

where $\bar{y}_i$ is the distance from the datum to the center of gravity of a subcomponent, A_i is the area of a subcomponent and N is the number of subcomponents as shown in Figure 7.11. The process can be repeated in the x direction. A void or hole in an object can be treated as a subcomponent with negative area.

Example 7.8.1. Find the location of the centroid of the two-dimensional object shown in Figure 7.12 relative to datums on the bottom and left sides of the object.

Solution. Tables containing the relevant terms, such as in Table 7.2, are convenient means of solving this type of problem.

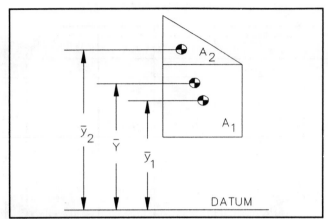

Figure 7.11 Center of Gravity of a Composite Object.

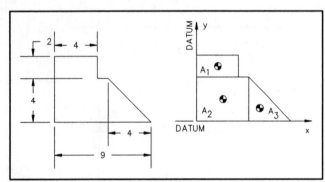

Figure 7.12 Composite Object.

Table 7.2 Centroid of Composite Object

i	$\bar{x}_i$	$\bar{y}_i$	A_i	$\bar{x}_i A_i$	$\bar{y}_i A_i$
1	2.00	5.00	8.00	16.00	40.00
2	2.50	2.00	20.00	50.00	40.00
3	6.33	1.33	8.00	50.64	10.64
Σ			36.00	116.64	90.64

$$\bar{X} = \frac{116.64}{36.00} = 3.24 \qquad \bar{Y} = \frac{90.64}{36.00} = 2.52$$

7.9 Friction

The force required to overcome friction resulting from bodies in contact is important in many statics problems. In almost all cases, the concept of *dry* or *Coulomb* friction is assumed to apply. The force of friction acts opposite to the direction of motion that would result from an applied force. The maximum possible force of friction is defined as:

$$F_F = \mu N$$

where μ is the coefficient of friction and N is the force that acts normal to the surfaces in contact. It should be noted that this discussion applies to *static* friction. A similar relationship exists for *dynamic* friction that exists between bodies in relative motion. Also, the coefficient of friction is a function of the two surfaces in contact. To overcome friction and cause a body to move, a force F must be applied that is greater than or equal to the force of friction as illustrated in Figure 7.13.

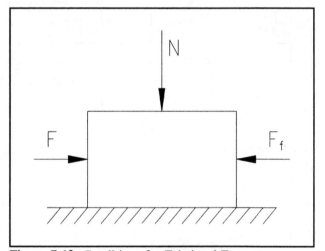

Figure 7.13 Conditions for Frictional Forces.

Example 7.9.1. The block shown in Figure 7.13 has a mass of 200 kg and the coefficient of friction between the block and the floor is 0.3. How large must the force F_F be to start the block in motion?

Solution. The normal force acting on the frictional contact surfaces is the weight of the block. The weight is given by:

$$W = mg = (200\ kg)(9.8\ m/s^2) = 1960 N$$

The force F must be at least equal to the force of friction:

$$F \geq F_F = \mu N = (0.3)\ 1960 N = 588 N$$

Practice Problems

7.1 Find the reaction at the right side of the beam shown in Figure P7.1.

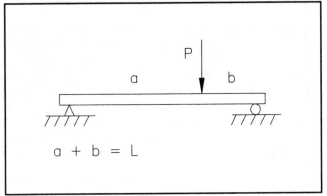

a + b = L

Figure P7.1 Problem 7.1.

7.2 A pulley is mounted at the end of a beam, as shown in Figure P7.2. The pulley weighs 20 pounds. The load suspended on the cable weighs 40 pounds. Find the force the beam exerts on the pulley.

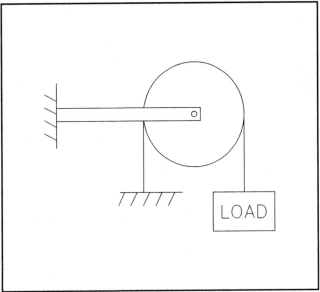

Figure P7.2 Problem 7.2.

7.3 Find the tension in the cable supporting the beam shown in Figure P7.3.

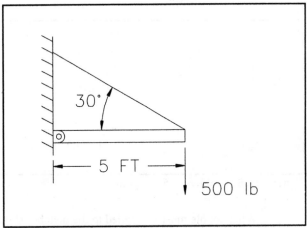

Figure P7.3 Problem 7.3.

7.4 A ladder is leaning against a wall as shown in Figure P7.4. It has a uniform cross section and a mass of 50 kg. It is leaning against a smooth (frictionless) wall with its lower end on a level floor. A stop on the floor prevents the ladder from slipping. Find the force the wall exerts on the ladder.

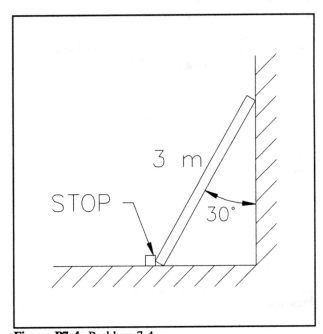

Figure P7.4 Problem 7.4.

7.5 A beam is loaded as shown in Figure P7.5. Find the reaction at the wall.

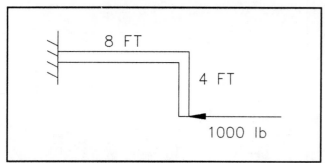

Figure P7.5 Problem 7.5.

7.6 What couple must be applied to the member shown in Figure P7.6 to keep it in equilibrium?

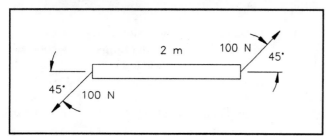

Figure P7.6 Problem 7.6.

7.7 The coefficient of friction between the box and the ramp shown in Figure P7.7 is 0.25. Will the box slide down the ramp?

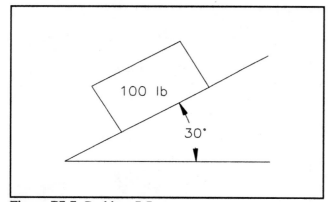

Figure P7.7 Problem 7.7.

7.8 A box shown in Figure P7.8 is resting on a ramp with a coefficient of friction equal to 0.3. What is the magnitude of the force F that will prevent the box from sliding down the ramp?

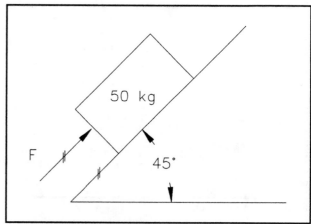

Figure P7.8 Problem 7.8.

7.9 Find the center of gravity of the object shown in Figure P7.9.

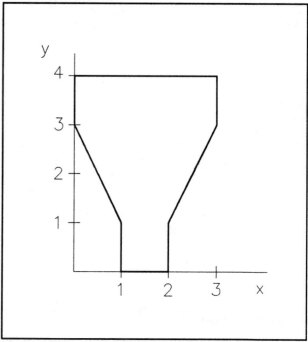

Figure P7.9 Problem 7.9.

7.10   Find the center of gravity of the object shown in
        Figure P7.10.

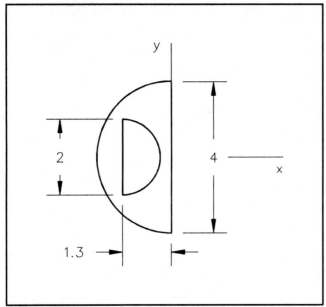

Figure P7.10 Problem 7.10.

DYNAMICS

Dynamics is the study of bodies in motion. There are two major topics in the study of dynamics:

- **Kinematics,** the study of the *motion* of particles and bodies, and

- **Kinetics,** the study of the *forces and moments* required to induce motion.

The motion of particles can be categorized as rectilinear and curvilinear. These types of motion also apply to rigid bodies. The motion of rigid bodies can also be described by angular motion.

8.1 Rectilinear Motion

Rectilinear motion describes the action of a particle in a straight line. The acceleration, *a*, velocity, *v* and displacement, *s* of a particle are described by the following relationships:

- The **average velocity** in a time interval, Δt, is expressed as:

$$v_{AVG} = \frac{\Delta s}{\Delta t}$$

- **Instantaneous velocity** is given by the derivative relationship:

$$v = \frac{ds}{dt}$$

- The **average acceleration** in a time interval, Δt, is expressed as:

$$a_{AVG} = \frac{\Delta v}{\Delta t}$$

- **Instantaneous acceleration** is given by:

$$a = \frac{dv}{dt}$$

For systems with constant acceleration (such as the action of gravity), integration of the derivative definition of acceleration can be performed with initial conditions at $t = 0$, $v = v_0$ and $s_0 = 0$ to define a set of useful relationships:

$$v = v_0 + at$$

$$s = v_0 t + \frac{at^2}{2}$$

$$v^2 = v_0^2 + 2as$$

Example 8.1.1. An automobile skids to a stop in 200 feet after its brakes are applied when it was moving at 60 miles/hour. Find the acceleration in units of ft/s^2, assuming the deceleration is constant.

Solution. The initial velocity must be put in appropriate units:

$$v_0 = \frac{60\, miles}{hour} \cdot \frac{hour}{3600\, s} \cdot \frac{5280\, ft}{mile} = 88 \frac{ft}{s}$$

The following equation of rectilinear motion will be applied:

$$v^2 = v_0^2 + 2as$$

If the final velocity is taken as zero, this equation can be algebraically rearranged to yield:

$$a = -\frac{v_0^2}{2s} = -\frac{88^2}{(2)(200)} = -19.4\, ft/s^2$$

The negative sign indicates that the vehicle is decelerating.

67

8.2 Curvilinear Motion

Curvilinear motion describes the action of particle travelling in a plane curve. A plane curve may be approximated over a small interval with a circular arc with radius of curvature, r. The motion is characterized by components along the normal n and tangent t to the curve at the instantaneous position of the particle as seen in Figure 8.1. The acceleration vector of the particle can be described as the sum of two components along the normal and tangential directions:

$$a_n = \frac{v^2}{r}$$

$$a_t = \frac{dv}{dt} \text{ or } \frac{\Delta v}{\Delta t}$$

The tangential acceleration characterizes the change in the speed of the particle over time and the normal acceleration characterizes the change in direction over time.

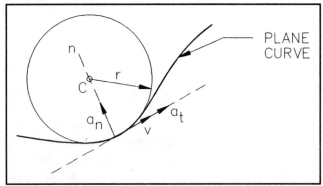

Figure 8.1 Curvilinear Motion.

Example 8.2.1 A train is traveling at 80 km/hour around a curve with a radius of curvature of 1000 meters. The brakes are applied and the train slows at a constant rate. The brakes are released after five seconds, resulting in the train now moving at 20 km/hour. Find the acceleration at the instant of braking in m/s^2.

Solution. The speeds should be expressed in the appropriate units:

$$80\,\frac{km}{hour} \cdot \frac{1000\,m}{km} \cdot \frac{hour}{3600\,s} = 22\,\frac{m}{s}$$

$$20\,\frac{km}{hour} = 5.6\,\frac{m}{s}$$

The tangential component of the acceleration will be approximated by the finite change in speed over the time interval:

$$a_t = \frac{5.6\,m/s - 22.2\,m/s}{5s} = -3.3\,\frac{m}{s^2}$$

The normal component of the acceleration at the instant braking began is given by:

$$a_n = \frac{v^2}{r} = \frac{(22.2\,m/s)^2}{1000\,m} = 0.493\,\frac{m}{s^2}$$

8.3 Angular Motion

A rigid body can be characterized by angular motion where the angular displacement of the body about a point is measured relative to a datum (usually the positive x axis). Figure 8.2 shows a body rotating with angular

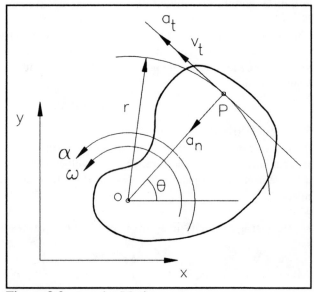

Figure 8.2 Angular Motion.

displacement θ, angular velocity ω and angular acceleration α. These terms are related by:

$$\omega = \frac{d\theta}{dt}$$

$$\alpha = \frac{d\omega}{dt} = \frac{d^2\theta}{dt^2}$$

The tangential velocity and acceleration of point P are given by:

$$v_t = r\omega \qquad a_t = r\alpha$$

The normal acceleration of point P is given by:

$$a_n = \frac{v_t^2}{r} = r\omega^2$$

If the angular acceleration α is constant, $\omega = \omega_0$ and $\theta_0 = 0$ at $t=0$, then the following equations of motion will apply:

$$\omega = \omega_0 + \alpha t$$

$$\theta = \omega_0 t + \frac{\alpha t^2}{2}$$

$$\omega^2 = \omega_0^2 + 2\alpha\theta$$

8.4 Newton's Second Law

Newton's Second Law describes the relationship between the forces acting on a particle or body and how it will accelerate:

$$\Sigma F = ma$$

The mass of the body is assumed to be constant and a is the acceleration of the centroid (or center of mass) of the object. If the object is in rotational motion in a plane, the relationship between the moments acting on the body and the angular acceleration is given by:

$$\Sigma M = I\alpha$$

where I is the mass moment of inertia. The moments must be summed about the centroid. The mass moment of inertia for various bodies is shown in Figure 8.5.

Note: When the mass of an object is required for an object, it must be expressed in the proper units. In the metric system, mass is measured in kilograms. In the United States Customary System (USCS) or English system, mass, m, is measured in $lb \cdot s^2/ft$ or *slugs* can be found by:

$$m = \frac{W}{g}$$

where W is the weight in pounds and g is gravitational acceleration ($32.2\ ft/s^2$).

Example 8.4.1. Find the acceleration of the block shown in Figure 8.3. The coefficient of friction, μ, is 0.3.
Solution. The frictional resistance to motion must be determined. The normal force acting on the block is the sum of the applied force and the weight of the block:

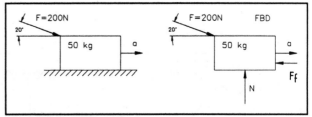

Figure 8.3 Sliding Block.

$$N = \sin 20°\,(200\,N) + (50\,kg)(9.8\,m/s^2) = 558\,N$$

The force of friction is given by:

$$F_F = 0.3(558\,N) = 167\,N$$

The forces acting in the horizontal direction are summed and Newton's Second Law is applied:

$$\Sigma F_H = ma = \cos 20°\,(200\,N) - 167\,N = (50\,kg)a$$

$$a = 0.4\,m/s^2$$

Example 8.4.2. A force of 10 N is applied to the cable wrapped around the 50-kg pulley shown in Figure 8.4. What is the angular acceleration of the pulley?

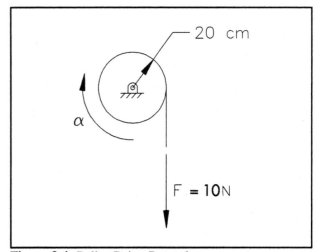

Figure 8.4 Pulley Being Rotated.

Solution. The mass moment of inertia of the pulley is determined using Figure 8.5. Assuming that the pulley can be approximated as a disk, I is given by:

Shape	(Origin at the center of mass.)	
Slender Rod	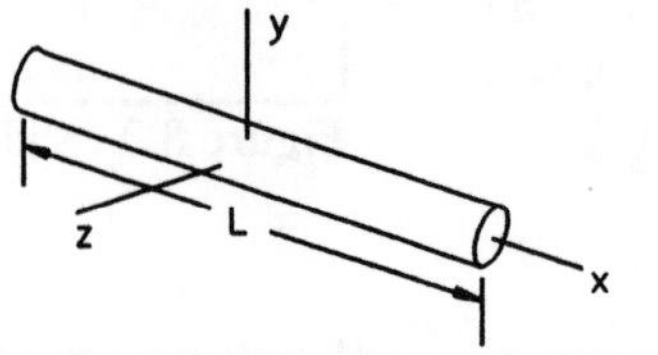	$I_x = I_y = 1/12\ mL^2$
Thin Rectangular Plate	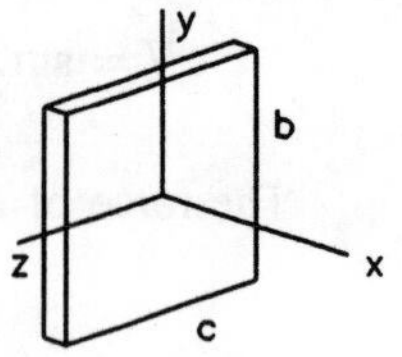	$I_x = 1/12\ m(b^2+c^2)$ $I_y = 1/12\ mc^2$ $I_z = 1/12\ mb^2$
Rectangular Prism	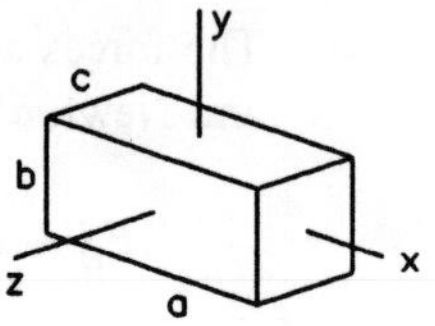	$I_x = 1/12\ m(b^2+c^2)$ $I_y = 1/12\ m(c^2+a^2)$ $I_z = 1/12\ m(a^2+b^2)$
Thin Disk	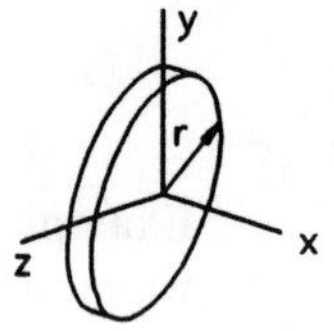	$I_x = 1/2\ mr^2$ $I_y = I_z = 1/4\ mr^2$
Circular Cylinder	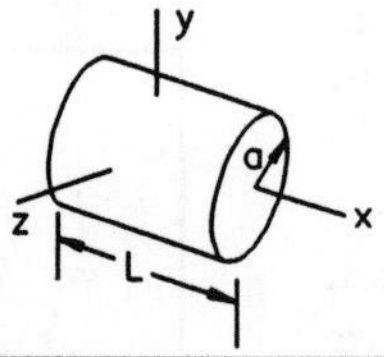	$I_x = 1/2\ ma^2$ $I_y = I_z = 1/12\ m(3a^2+L^2)$

Figure 8.5 Moment of Inertia of Common Shapes.

$$I = mr^2/2 = (50\,kg)(0.2m)^2/2 = 50\,kg \cdot m^2$$

Summing moments about the center of the pulley yields:

$$\Sigma M_c = I\alpha$$

$$(10\,N)(0.2m) = (50\,kg\,m^2)\alpha$$

$$\alpha = 0.04\,rad/s^2 \;\; clockwise$$

8.5 Energy Methods

Energy methods are important tools for solving kinetics problems that would be cumbersome to solve by application of Newton's Laws. These techniques use the concepts of *conservation of energy* (energy can neither be created or destroyed) and the definition of *work* (energy is the capability to do work).

Work is defined as the product of an applied force, F, and the distance over which the force is applied, s. For a constant force, this relation is given by:

$$W = F \cdot s$$

For a rotating body, work is the product of an applied moment, M, and the angle, θ, (in radians) through which the moment is applied. The net work done on an object is equal to the change in energy in the object. The energy change can occur in several ways:

- *Kinetic Energy*. For a body in linear motion, this is given by:

$$KE = \frac{1}{2}mv^2$$

For a body in angular motion, this is given by:

$$KE = \frac{1}{2}I\omega^2$$

- *Potential Energy*. This is the stored energy associated with the body or the *potential* to do work. One form of potential energy is the position of an object relative to a datum in a gravitational field. The potential energy is given by the product of the weight of the object, mg, and its distance from the selected datum, h:

$$PE = mgh$$

In the case of the energy stored in a linear spring, the potential energy is given by:

$$PE = \frac{1}{2}kx^2$$

where k is the spring constant and x is the distance that the spring is compressed or extended.

Example 8.5.1. Find the angular velocity of the 20-kg cylindrical pulley after the 30-kg block falls 1.5 meters from rest, as shown in Figure 8.6. Neglect bearing friction in the pulley.

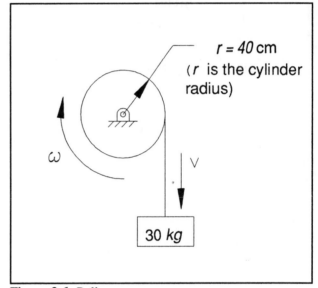

$r = 40\,cm$
(r is the cylinder radius)

Figure 8.6 Pulley.

Solution. The work done by gravity is equal to the kinetic energy of the pulley and the block:

$$W = mgh = \frac{1}{2}mv^2 + \frac{1}{2}I\omega^2$$

Using the definition of the moment of inertia of a cylinder and tangential velocity:

$$m_{block}gh = \frac{1}{2}m_{block}(\omega r)^2 + \frac{1}{2}(m_{pulley}r^2)\omega^2$$

$$(30)(9.8)(1.5) = \frac{1}{2}(30)(0.4\omega)^2 + \frac{1}{2}(\frac{1}{2}(20)(0.4)^2)\omega^2$$

$$\omega = 10.2\,rad/s \;\; clockwise$$

Example 8.5.2. Find the velocity of the 50-kg block shown in Figure 8.7 after it falls 2 meters from rest.

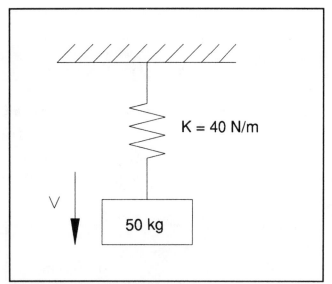

Figure 8.7 Mass/Spring System.

Solution. The work and energy method is applied with the work done by block being equal to the kinetic energy of the block and potential energy stored in the spring:

$$mgh = \frac{1}{2}mv^2 + \frac{1}{2}kx^2$$

$$(50)(9.8)2 = \frac{1}{2}(50)v^2 + \frac{1}{2}(40)(2)^2$$

$$v = 6 \ m/s$$

8.6 Momentum

Momentum is a description of the kinetic state of a body. *Linear momentum* is given by the product of mass and linear velocity:

$$p = mv$$

Similarly, *angular momentum* is the product of moment of inertia and angular velocity:

$$H = I\omega$$

If the resultant force or moment acting on an object is zero, the momentum of the object remains constant in magnitude and direction. If the mass remains constant, the relationships between force or moment and changing momentum are:

$$F\Delta t = m\Delta v$$

$$M\Delta t = I\omega$$

The product of force or moment and a time interval, Δt, is known as an *impulse*. The momentum of the system after the application of a force for a period of time is given by the sum of the original momentum of the system and the impulse:

$$mv_1 + F\Delta t = mv_2$$

$$I\omega_1 + M\Delta t = I\omega_2$$

Example 8.6.1. A 2500-pound car moving at 60 miles/hour is stopped by applying brakes which exert a 1500-pound braking force at the tires. How long must the brakes be applied to bring the vehicle to a stop?

Solution. The method of impulse and momentum will be applied noting that the impulse is in the direction opposite the motion of the vehicle, as illustrated in Figure 8.8.

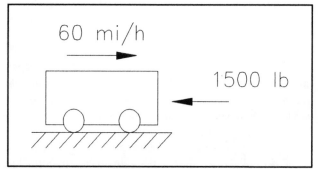

Figure 8.8 Car in Braking.

First the initial velocity is converted into consistent units:

$$60 \ miles/hour = 88 \ ft/s$$

The duration of the impulse can be found by observing that the final velocity is zero:

$$mv_1 + F\Delta t = mv_2$$

$$(2500/32.2)(88) - 1500\Delta t = 0$$

$$\Delta t = 4.6 \ s$$

Practice Problems

8.1   An object is moving with an initial velocity of 30 *m/s*. If it is decelerating at 5 *m/s²*, how far will it travel before it stops?

8.2   A particle is shot straight up with an initial velocity of 50 *m/s*. After how many seconds will it return if the drag is neglected?

8.3   A car is travelling at 25 *m/s*. It takes 0.3 seconds to stop the car with the brakes applying a deceleration of 6 *m/s²*. How far will the car travel before it stops?

8.4   A gear is accelerated at 6 *rad/s²*. How many times will the gear revolve in ten seconds?

8.5   A 1.5-foot-long lever is hinged at one end and is rotating at 0.5 *rad/s*. It is accelerating at 1 *rad/s²*. How many seconds will it take for the free end to reach 10 *ft/s?*

8.6   A car turns a corner at 40 *miles/hour*. The corner has a radius of curvature of 200 ft. Calculate the acceleration in *ft/s²*.

8.7   A wheel is 2 feet in diameter. It rolls without slipping on a flat surface at 10 *ft/s*. What is the angular velocity of the wheel?

8.8   A block of weight, *W*, is on a 60-degree incline (with respect to horizontal). The coefficient of friction between the block and the incline is 0.3. How far will the block move in five seconds if it is released from rest?

8.9   A weight is suspended by two cables, AB and BC as shown in Figure P8.1. What is the ratio of the tensions in cable AB just before and just after cable AC is cut?

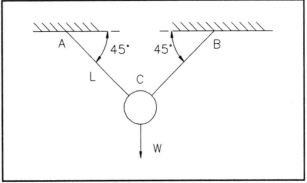

Figure P8.1   Problem 8.9.

8.10   A weight *W* will cause a spring to deflect 1 inch if it rests on top of it. If the weight is dropped from a height of 10 inches above the free position of a spring, how much will the spring deflect?

8.11   A 100-pound missile strikes a concrete wall with an impact velocity of 2000 *ft/s*. The projectile will come to rest in 0.001 s, after penetrating the wall 2 feet. Find the average force exerted on the wall.

8.12   The cylindrical pulley shown in Figure P8.2 has a mass of 10 kg. Find its angular acceleration.

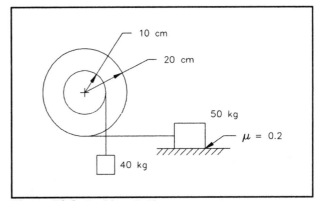

Figure P8.2   Problem 8.12.

STRENGTH OF MATERIALS

Strength of materials is the study of deformable bodies subject to applied forces and moments. Some of the important issues in the study of the strength of materials are:

• *How much load can be safely applied to a structure or component?*

• *What material should be chosen to fabricate a component to safely withstand a particular load?*

• *How much will a component deflect under load?*

9.1 Stress and Strain

Stress and strain are quantities that are used to characterize the strength and deformation of a component. When the properties of an engineering material are tested and recorded in a handbook, the definition of stress and strain allow the test data to be applied to virtually any structure.

In a mechanical strength test of a material such as steel, a test specimen is loaded under with a controlled amount of force, applied perpendicular to the cross sectional area, and the amount of resulting deformation is recorded until it fractures. Figure 9.1 shows a prismatic (constant cross section) bar of length L and cross sectional area A is subjected to a axial force F. The applied force causes the bar to stretch by ΔL.

The normal or longitudinal strain ϵ is defined as:

$$\epsilon = \frac{\Delta L}{L}$$

Strains are positive if the specimen is elongated and negative if it is shortened. The *normal or axial stress* σ is defined as:

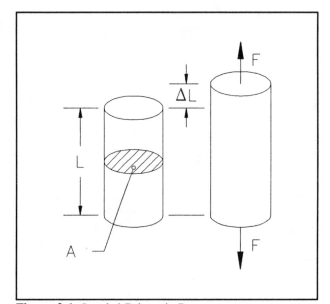

Figure 9.1 Loaded Prismatic Bar.

$$\sigma = \frac{F}{A_0}$$

Stress has units of force per unit area. Common stress units are the *Pascal* or Pa (defined as one Newton per square meter) and psi (an abbreviation for pounds per square inch). Since the Pascal and the psi are relatively small, stress is usually expressed in megapascals (mPa) and thousands of psi (kpsi).

9.2 Axial Loading

If an object is subjected to a positive strain in one direction, it is normal for the object to contract or experience a negative strain in another direction. Figure 9.2 shows a bar being strained in the x direction due to the application of a force along that axis. A negative strain is also induced in the y and z directions. An important material property is defined for the ratio of strains in two directions in a specimen. For an isotropic material (where mechanical properties are the same in all directions), *Poisson's Ratio* ν is defined as:

$$\nu = -\frac{\epsilon_y}{\epsilon_x} = -\frac{\epsilon_z}{\epsilon_x}$$

for the bar shown in Figure 9.2.

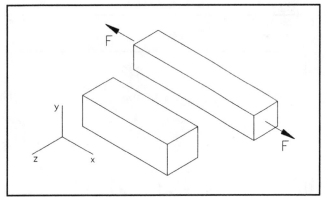

Figure 9.2 Axial and Lateral Strain.

If the specimen shown in Figure 9.1 is subjected to a *tensile test*, an increasing axial force will be applied and the resulting deformation will be recorded. This data is transformed into stress and strain and plotted to depict important material properties that characterize the strength of the material. A typical stress/strain curve resulting from a tensile test of a *ductile material* (capable of withstanding significant strain prior to fracture) is shown in Figure 9.3. Steel, aluminum and brass are some examples of ductile materials.

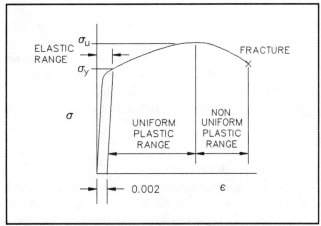

Figure 9.3 Stress-Strain Diagram.

As the force is initially applied, the stress increases proportionally with strain. If the material is a *linear elastic material*, strain and stress in the elastic range will be related by Hooke's Law:

$$\sigma = E \epsilon$$

where *E* is a material constant called *Young's modulus or modulus of elasticity*. The component will return to its original undeformed shape if the load is removed in the elastic range. Clearly, most mechanical design is done for the elastic range. As the material continues to deform under increasing applied load, the elastic behavior will cease and the material will take on a permanent set if the load is released. This phenomenon is known as *plastic deformation*. The stress corresponding to the transition from the elastic region to the plastic region is called the *yield strength*. For consistent test results, the yield point is defined as where a line drawn parallel to the elastic region with a 0.002 strain offset intersects the stress/strain curve. The transition point between the elastic and plastic regions is also called the *proportional limit*.

Once the material passes the yield strength, it will undergo uniform plastic deformation as seen in Figure 9.4(b) if the load is continually increased. The

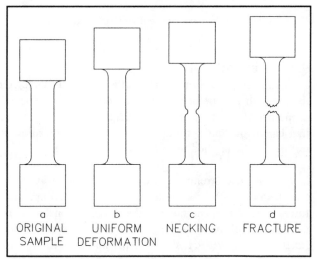

Figure 9.4 Tensile Test Sample.

applied load necessary to elongate, σμ, the sample will increase until the *ultimate strength* is reached. At this point, the sample *necks* (deforms nonuniformly) as in Figure 9.4(c). If loading is sustained, the sample will elongate until fracture. The significant material properties that can be collected from a tensile test are shown for various materials in Table 9.1. An additional property which can be defined, called the *shear modulus* or *modulus of rigidity* is defined as:

$$G = \frac{E}{2(1+\nu)}$$

This property is important for analyzing *shear stress*. Shear stress results from a force applied parallel to a cross sectional area.

The definitions of stress and strain can be combined to derive a useful formula for the amount of elongation that occurs in a tensile member when axially loaded in the elastic region:

$$\Delta L = \frac{F L}{A E}$$

The *factor of safety*, **FS**, in the design of a component is the ratio of the allowable stress to the actual stress:

$$FS = \frac{\sigma_{allowable}}{\sigma_{actual}}$$

As stated earlier, the maximum allowable stress is typically the yield stress for the material.

Table 9.1 Typical Material Properties.

Material	Modulus of Elasticity E 10^6 psi	Modulus of Rigidity G 10^6 psi	Ultimate Strength 10^3 psi	Yield Strength 10^3 psi	Poisson's Ratio
Mild Steel	30	12	58	36	.3
Aluminum	10	3.9	16	14	.33
Magnesium	6.5	2.4	55	40	.35
Titanium	15	6	130	120	.34
Brass	15	6	48	15	.33

Note: 1. Units of psi can be approximately converted to kPa by multiplying by 7000.
2. This data is for example purposes only. It should not be used for design.

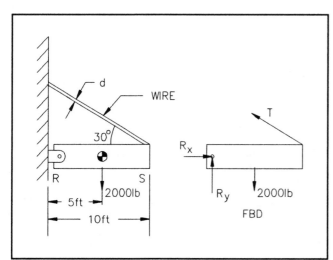

Figure 9.5 Hinged Beam.

Example 9.2.1. A hinged beam weighs 2000 pounds and is supported by a single steel wire shown in Figure 9.5. Using the material property data from Table 9.1 and a factor of safety of two, find the required diameter of the wire and the amount of elongation in the wire.
Solution. The structure is reduced to a free body diagram and the tensile in the wire is found by summing moments about the wall connection, R:

$$\Sigma M_R = 0; \quad -2000(5)+(T\sin 30°)(10) = 0$$

$$F = T = 2000lb$$

The allowable stress is found by applying the factor of safety to the yield strength of steel:

$$\sigma_{allow} = \tfrac{1}{2}(36000) = 18000\,psi$$

The allowable stress is used to size the wire based on the normal stress formula:

$$\sigma_{allow} = \frac{F}{A} = \frac{2000\ lb}{\left(\dfrac{\pi d^2}{4}\right) in.^2} = 18000$$

$$d = 0.38\ in.$$

The elongation can be readily found since the factor of safety assures loading in the elastic range:

$$\Delta L = \frac{F\,L}{A\,E} = \frac{(2000)\left(\dfrac{120}{\cos 30^\circ}\right)}{\left(\dfrac{\pi(0.38)^2}{4}\right)30\times 10^6} = 0.08\ in$$

9.3 Torsional Loading

Shafts and other machine elements that are subjected to equilibrating couples at each end (torque) are in *torsion*. A circular shaft in torsion is shown in Figure 9.6. The applied

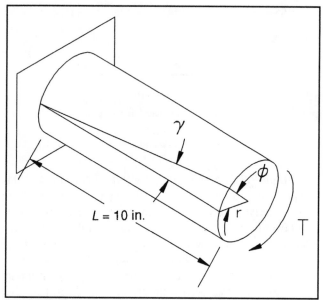

Figure 9.6 Shaft Loaded in Torsion.

torque creates a shear stress that causes the shaft to twist. In the elastic range of the material, the shear stress τ is related to the shear strain γ a modified version of Hooke's Law:

$$\tau = G\,psi\cdot\gamma\ in./in.$$

The maximum shear stress in the shaft occurs at the surface and is given by:

$$\tau_{max} = \frac{T\ in.\text{-}lb\ r\ in.}{J\ in.^4}$$

where T is the applied torque, r is the radius and J is the polar moment of inertia. The polar moment of inertia for a circular cross section is given by:

$$J = \frac{\pi\,r^4}{2}$$

The polar moment of inertia for a hollow shaft can be found by adding a negative polar moment corresponding to the void space: $J = \pi\left(\dfrac{R^4 - r^4}{2}\right)$

The angle of twist ϕ, with units in radians, for an elastically loaded shaft can be found by:

$$\phi = \frac{T\ in.lb\ L\ in.}{J\ in.^4 G\ lb/in^2}$$

Example 9.3.1. A circular steel rod supports a handle as shown in Figure 9.7. The length of the shaft L is 10 inches. Find the maximum shear stress in the rod and the angle of twist of the rod.

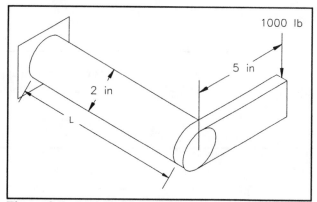

Figure 9.7 Steel Rod in Torsion.

Solution. The torque in the rod is given by:

$$T = (5)(1000) = 5000\ in\cdot lb$$

The maximum shear stress occurs at the surface of the rod and is given by:

$$\tau_{max} = \frac{T r}{J} = \frac{(5000)(1)}{\left(\dfrac{\pi(1)^4}{2}\right)} = 3200\,psi$$

The angle of twist is found by:

$$\phi = \frac{TL}{JG} = \frac{(5000)(10)}{\left(\frac{\pi(1)^4}{2}\right)(12 \times 10^6)} = 0.0026 \, rad = 0.15°$$

9.4 Beam Loading

Beams are machine elements that are typically much longer than they are wide and are loaded in a direction that is perpendicular to their long dimension. Beams may support any combination of loads, including their own weight. The two main issues in the analysis of beams are strength and the deflection. The two typical conditions for the loading of a beam are shown in Figure 9.8. Figure 9.8(a) shows a *cantilever beam* which has a fixed support at one end and Figure 9.8(b) shows a *simply supported beam* that has a pin support at one end and a roller support at the other (the roller support is assumed to simplify the analysis by avoiding longitudinal compression of the beam).

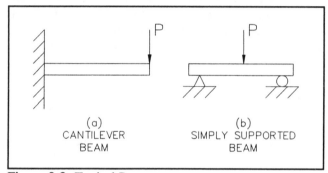

Figure 9.8 Typical Beams.

The typical beam loading conditions are schematized in Figure 9.9. Figure 9.9(A) and Figure 9.9(D) show *concentrated loads* where a load *P* in pounds or Newtons is applied at one point along the beam. Figure 9.9(B), Figure 9.9(C), and Figure 9.9(E) show uniformly *distributed loads* where the load *w* is expressed in pounds/inch or Newtons/meter. A heavy concrete beam, loaded by its own weight, is a good example of a distributed load.

In the cross section of a beam, an internal force *V* referred to as the *shear force* and an internal moment *M*, referred to as the *bending moment*, cause the beam to be stressed under loading. The shear force and bending moment can be determined by using a free body diagram of a loaded beam that has been segmented by making imaginary cuts between each load or reaction as shown in Figure 9.10. The maximum value or the shear force and bending moment for

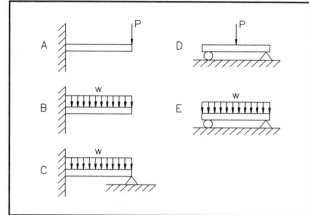

Figure 9.9 Typical Beam Loading Conditions.

a beam must be determined to analyze a design for safety.

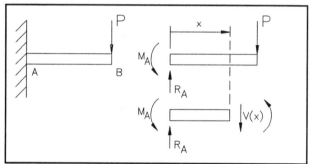

Figure 9.10 Determining Shear and Moment in a Beam.

Example 9.4.1. Find the distribution of the shear force and bending moment of the simply supported beam with a concentrated load shown in Figure 9.11(a). Determine the magnitude of the maximum shear force and bending moment.

Solution. A free body diagram of the beam is drawn as shown in Figure 9.11(b). A coordinate axis is imposed on the beam with the origin on the left side. Imaginary cuts are made in between the load and the reactions as shown in Figure 9.11(c). Individual free body diagrams are used for the isolated sections to determine the magnitude of the shear force and bending moment in between the load and the reactions. The resulting shear distribution is shown in Figure 9.11(d). The resulting bending moment distribution is shown in Figure 9.11(e).

The maximum values of shear force and bending moment for the common loading cases shown in Figure 9.9 are compiled in Table 9.2. In all cases, the length of the beam is **L**.

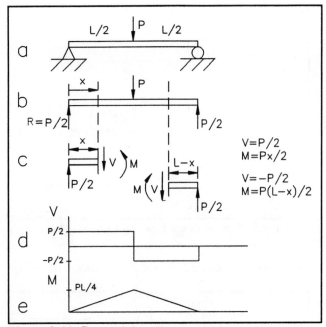

Figure 9.11 Determining Shear and Moment Diagrams.

Table 9.2 Important Beam Formulas.

Loading Condition	Maximum Shear Force	Maximum Bending Moment
A	P	PL
B	wL	$wL^2/2$
C	5wL/8	$wl^2/8$
D	P/2	PL/4
E	wL/2	$wL^2/8$

The loading of the beam induces two stresses: a bending stress σ that causes the beam to change shape and a shearing stress τ that causes the layers of internal microstructure of the beam to attempt to slide away from each other in the manner of a fanned-out deck of cards. Figure 9.12 shows a rectangular cross section cantilever beam with a concentrated load. The bending stress acts normal to the cross section and is tensile or positive on the top of the beam and negative on the bottom. The bending stress is zero on a plane that passes through the centroid of the cross section known as the *neutral axis*. The shear stress in a rectangular cross section is parabolically

distributed and takes a maximum value along the neutral axis.

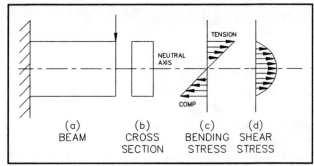

Figure 9.12 Stress Distribution in a Beam Cross Section.

The maximum bending stress is found by:

$$\sigma = \frac{Mc}{I}$$

where c is the distance from the neutral axis to the plane where the bending stress is to be calculated (this is typically the top or bottom surface) and I is the area moment of inertia. The area moment of inertia for a rectangular cross section is given by:

$$I = \frac{bh^3}{12}$$

where b is the base, h is the height and the load is applied in a direction parallel to the height.

The maximum value of shear stress for a narrow rectangular cross section beam is given by:

$$\tau = \frac{3 V_{max}}{2 A}$$

where V is the shear force and A is the cross sectional area of the beam.

Example 9.4.2. Find the maximum bending stress and shear stress in the beam shown in Figure 9.13.

Solution. The maximum values for shear force and bending moment were found in the previous example.

$$V_{max} = \frac{P}{2} = \frac{5000}{2} = 2500 \ N$$

$$M_{max} = \frac{PL}{4} = \frac{(5000)(4)}{4} = 5000 \ N{\cdot}m$$

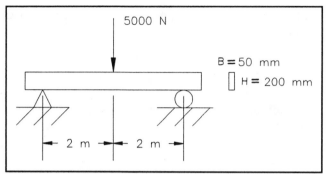

Figure 9.13 Simply Supported Beam.

The area moment of inertia of the cross section is found as:

$$I = \frac{b h^3}{12} = \frac{1}{12}(0.05)(0.2)^3 = 3.33 \times 10^{-5} \ m^4$$

The maximum value of shear stress is given by:

$$\tau_{max} = \frac{3 V}{2 A} = \frac{3(2500)}{2(0.05)(0.2)} = 0.375 \ MPa$$

The maximum value of bending stress is given by:

$$\sigma_{max} = \frac{M c}{I} = \frac{(5000)(0.1)}{3.33 \times 10^{-5}} = 15 \ MPa$$

The vertical displacement of the neutral axis of the beam is a measure of the beam deflection. Proper design of a beam must often consider deflection. Even though a beam may be suitable under strength criteria, its deflection may be excessive for the application. Table 9.3 lists the maximum deflections for the same loading conditions described in Figure 9.9 and Table 9.2. (Refer to Figure 9.9 for geometry and loading conditions.)

Table 9.3 Important Beam Deflection Formulas.

Loading Condition	Maximum Deflection
a	$PL^3/(3EI)$
b	$wL^4/(8EI)$
c	$wL^4/(185EI)$
d	$PL^3/(48EI)$
e	$5wL^4/(384EI)$

9.5 Column Loading

A column is a long slender member that is loaded axially in compression. Columns often fail by *buckling* rather than by yielding. Buckling is the elastic failure of an element where excessive deflection occurs. One can visualize buckling by holding a yardstick between his/her hands and loading it compressively until a sudden change in geometry occurs. Failure is clearly elastic, since the yardstick regains its original shape after the load is removed.

Column theory was derived for an ideal column that has two pinned end connections and special consideration given to assure pure compressive loading. The ideal column, known as a pinned-pinned ended column, is depicted in Figure 9.14(a). The characteristic change in geometry that occurs upon buckling shown in Figure 9.14(b).

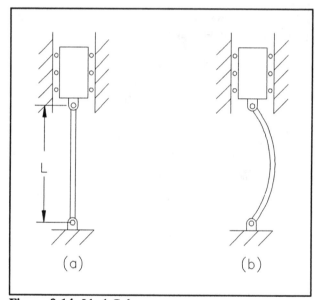

Figure 9.14 Ideal Column.

Buckling can be predicted by calculating the critical load at which buckling will occur. For the pinned-pinned ended column, the critical loaded can be predicted by:

$$P_{cr} = \frac{\pi^2 EI}{L^2}$$

This formula can be extended to more realistic loading conditions by observing that the some fraction of the buckled shape of the column is a pinned-pinned ended column. In this case, the critical load formula is applied with an *effective length, L_e*:

$$P_{cr} = \frac{\pi^2 EI}{L_e^2}$$

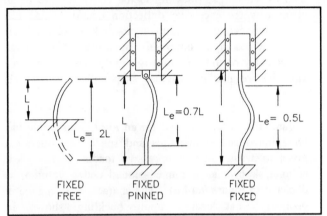

Figure 9.15 Effective Column Lengths.

Figure 9.15 shows three additional end conditions and the appropriate effective lengths.

Example 9.5.1. A 4-meter-long steel rod with square cross section (40 mm by 40 mm) is fixed at each end and loaded in compression. The modulus of elasticity is 210 GPa. The yield strength is 400 MPa. Find the critical load of the rod and determine if it will fail elastically.

Solution. The moment of inertia for the cross section can be found as:

$$I = \frac{bh^3}{12} = 2.13 \times 10^{-7} \, m^4$$

The effective length is based on a fixed-fixed type end condition:

$$L_e = 0.5L = 2 \, m$$

The critical load is given as:

$$P_{cr} = \frac{\pi^2 EI}{L_e^2} = \frac{(3.14)^2 (210 \times 10^9)(2.13 \times 10^{-7})}{2^2} = 110 \, kN$$

The critical load is used to determine the normal stress in the cross section. If the normal stress resulting from the critical load exceeds the yield stress, then the column will fail in yielding before it fails in buckling.

$$\sigma = \frac{F}{A} = \frac{110 \times 10^3}{0.04^2} = 68 \, kPa$$

The normal stress is far less than the yield stress, thus the column will fail in buckling.

Practice Problems

9.1 A 2000-pound load is supported by a 3/4-inch-diameter eyebolt. Find the stress in the straight section of the bolt.

9.2 The Poisson ratio of a material is 0.3. A cylinder made of this material is 30 mm in diameter and 150 mm long. If the cylinder increases in diameter by 0.0001 mm when it is compressed, how much does the length change?

9.3 A specimen is 4 inches long and 0.505 inch in diameter. It elongates elastically by 0.006 inch under a 12,000-pound load. Find the elastic modulus of the material.

9.4 The eyebolt in Problem 9.1 is made of mild steel. What is the factor of safety?

9.5 An outdoor sculpture is supported by a 50-foot-long mild steel wire. The wind can cause the sculpture to place a 200-pound load on the wire. The deflection of the sculpture is limited to 0.5 inch. What is the required diameter of the wire?

9.6 Four short table legs are made of 1-inch-diameter tubing with a 0.125-inch wall thickness. The table legs are made of typical aluminum. What is the maximum load that can be supported? The required factor of safety is four.

9.7 The diameter of a drive shaft is 0,375 inch. The maximum shear stress that can be placed on the shaft is 8000 psi Find the maximum transmissible torque. Use a factor of safety of two.

9.8 An extension on a socket wrench may be represented as a 0.5-inch-diameter steel cylinder 8 inches in length. A 75-foot/lb torque is applied to the wrench extension. How much does the extension twist?

9.9 A heavy concrete beam weighing 400 pounds/foot is simply supported. The beam is 8 feet long, 18 inches wide, and 24 inches tall. What is the bending stress in the beam?

9.10 A 1-inch by 12-inch floor joist 12 feet long is simply supported. A concentrated load of 350 pounds is placed in the center of the span. Find the maximum bending stress and shear stress.

9.11 A 0.5-inch-wide by 2-inch-tall steel bar is 36 inches long. It is clamped in a vise at one end. A 200-pound load is applied to the free end. Find the deflection.

9.12 A 60-inch-long steel bar has a cross section of 0.5 inch by 0.75 inch. It is clamped in a vise at one end. The other end is free. How much compressive load can be applied before the bar will buckle?

9.13 A square steel post is supporting an elevated platform. The post is 12 feet long and is welded to both the floor and the platform. The post must support a load of 250 pounds. Find the required cross section.

MATERIALS SCIENCE AND METALLURGY

Materials are a basic element of the engineering enterprise. The effective application of materials is critical to the success of the manufacturing of a product and its ultimate use. Metals remain the most important raw materials in engineering, although this situation is constantly being challenged by polymers and plastics. This section will review the basic characteristics of metals and their application.

10.1 Structure of Matter

The properties of metals are ultimately related to their atomic structure. The state of a metallic material is dependent on the type of atoms that it consists of and the forces between the atoms or *energy level*. The energy level of a metal determines the state or *phase* of that material. The basic phases of a material are solid, liquid and gas. Temperature and pressure can be used to define the phase as shown in Figure 10.1. At high energy levels, a material may exist in a gaseous or vapor stage. As atoms become less mobile with a decreasing energy level, the material may change to a liquid or solid phase. In a pure substance, there is a special condition of temperature and pressure where solid, liquid, and gas phases may exist simultaneously called the *triple point*.

Temperature has a significant role in determining the properties of a material. It is a way of characterizing the level of energy in an atom. The energy level determines the type of interaction occurring between an atom and the surrounding atoms. Consequently, temperature has a significant effect on the physical and chemical properties of a material. The energy of liquid is reduced by removing heat. The attraction between atoms increase until atoms order themselves into three-dimensional patterns or *crystals*. Since cooling is typically nonuniform, the first atoms to form crystals act as nuclei for further crystal formation. Other atoms join the solidifying mass and orderly growth continues until crystals encounter obstructions from other crystal growth

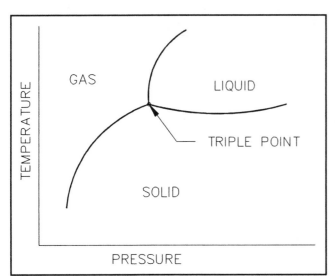

Figure 10.1 Phase Diagram.

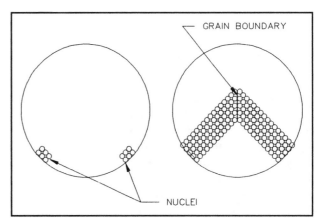

Figure 10.2 Grain Formation.

85

patterns from other nuclei as shown in Figure 10.2. These obstructions form the *grain boundaries* that compose the microscopically visible structure of the solid.

Individual *grains* are the structure that determines many of the physical properties of the material. Two metallic specimens may have the same chemical composition, but they may exhibit radically different hardnesses because of the grain structure. In general, a material with a fine grain structure will be harder and less ductile than a material with a coarse grain structure. A metal specimen is deformed by the relative displacement of crystals. A material with a fine grain structure provides more obstructions to the sliding and relative displacement of crystals than a coarse structure. Consequently, the material does not deform as readily.

10.2 Phase Diagrams

The *phase diagram* is a tool for understanding the phase changes for a material. Pure metals have a clearly defined melting point. Solidification occurs at a constant temperature. As a pure liquid metal is cooled, its temperature drops to the freezing temperature and remains at that temperature until all of the liquid has solidified. Figure 10.3 illustrates the solidification cycle for a pure metal.

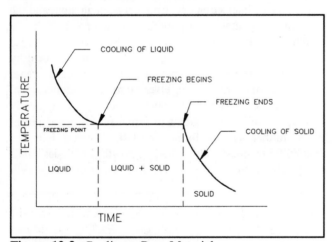

Figure 10.3 Cooling a Pure Material.

Alloys or mixtures of metals solidify over a range of temperatures based on the composition of the alloy. As an alloy is cooled, the mixture will begin to freeze at the *liquidus* temperature and completes freezing at the *solidus* temperature as shown in Figure 10.4. The mixtures will be in a slush state when it is cooling from the liquidus to solidus temperatures. The material illustrated in Figure 10.4 is an example of a *binary* alloy (two

components). This particular example is an alloy made of components which exhibit complete solid solubility.

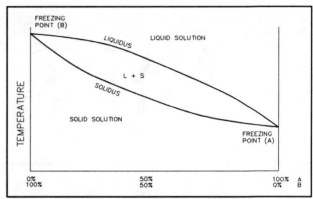

Figure 10.4 Phase Diagram of a Mixture.

Other binary alloys can have different phase diagrams where single phases of solid solutions can occur. Solid solutions can be visualized as a uniform distribution of two types of crystalline structures. A solid solution has all of the macroscopic properties of a solid. However, it is composed of two different internal arrangements of atoms in the form of crystals. Figure 10.5 shows a phase diagram for an alloy that exhibits a *eutectic point* (meaning "easy to melt"). A particular composition of the two components has a melting temperature that is lower than the melting temperature of either component. Lead-tin alloys such as solder make use of this characteristic.

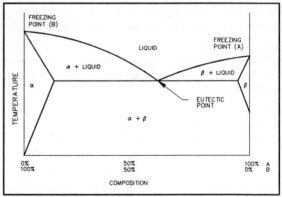

Figure 10.5 Phase Diagram for a Mixture with a Eutectic Point.

The phase diagram shown in Figure 10.5 has two single phase regions designated as α and β. The single phase regions make a transition into liquid with increasing temperature by first forming a two phase mixture with the liquid state.

10.3 Iron-Carbon Diagram

The most important phase diagram in engineering applications is the iron-carbon system. Steels, cast iron and cast steels are the most common engineering materials because of their versatile properties and relatively low cost. Virtually all iron products contain some carbon resulting from the production of the iron. Commercially used pure iron contains up to 0.008% carbon. Steels contain typically less than 1% carbon. Cast irons contain less than 6.7% carbon. The iron-carbon diagram is shown in Figure 10.6. It is only illustrated for carbon up to 6.7% carbon.

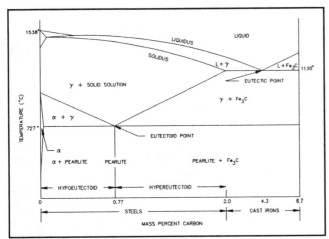

Figure 10.6 Iron-Carbon Diagram.

There are three solid phases of the iron-carbon mixture that are important in understanding the metallurgy of iron and steel. They are as follows:

• **Ferrite.** Ferrite, or α-iron, is a soft and ductile phase of iron. The solid solution contains only about 0.008% carbon at room temperature. It is the softest structure that is formed in the iron-carbon system. The typical hardness is approximately 80 BHN.

• **Austenite.** Austenite or γ-iron is a solid solution containing up to 2.11% carbon **1148°C (2098°F)**. Austenite is an important phase in the processing of steel. It is ductile at elevated temperatures and exhibits good formability. Steel can be *austenized* or transformed into a homogeneous structure of austenite by elevating its temperature according to the iron-carbon phase diagram. This is the usual starting point for virtually all heat treatment processes.

• **Cementite.** Cementite is 100% iron carbide (Fe_3C). It contains 6.67% carbon by weight. Cementite is a very hard and brittle structure that can have a significant influence on the properties of steels.

10.4 Heat Treatment of Steels

The important region of the iron-carbon diagram for steels is shown in Figure 10.7. Various microstructures can be formed by heating and controlled cooling of a steel specimen. For example, consider a sample of steel with a *eutectoid* composition (0.77% C) which is heated from room temperature past the austenizing temperature (from Figure 10.7, 727°C). The sample will transform into a uniform mass of γ-iron. If the sample is cooled slowly (to maintain equilibrium conditions), a phase transformation will occur when the temperature drops below 727°C. The austenite will be transformed into ferrite and cementite. At the eutectoid composition, the structure of ferrite and cementite is called *pearlite*. The structure of pearlite is characterized by alternating layers (*lamellae*) of crystals of ferrite and cementite (under approximately 2500x magnification, the structure appears like a fingerprint). Most significantly, the resulting structure has mechanical properties between ferrite (soft and ductile) and cementite (hard and brittle).

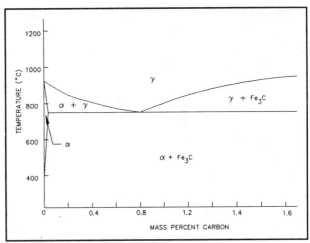

Figure 10.7 Iron-Carbon Diagram (Steels Only).

In the heat treatment of a sample of *hypoeutectoid* steel (with carbon contents less than 0.77%), the resulting structure consists of a mixture of ferrite and pearlite (giving a correspondingly softer and more ductile composition than pure pearlite). *Hypereutectoid* steels produce a structure of cementite and pearlite which is correspondingly more brittle and hard than pure pearlite. As can be seen in Figure 10.6, both hypoeutectoid and hypereutectoid steels have austenizing temperatures greater than 727°C.

If the crystal structure of the pearlite after heat treatment is thin and closely packed it is called *fine pearlite*. If the crystals are large and loosely packed, it is called *coarse pearlite.* Most heat treatment involves the formation of pearlite in a controlled way. The mechanical properties of the steel can be controlled by heating the steel into the austenizing range and cooling it at a rate that gives the desired pearlite structure. Large crystals take time to grow (as anyone who has tried to make rock candy has discovered). Consequently, coarse pearlite results from a relatively slow cooling process and fine pearlite results from a rapid cooling process. As described earlier, fine crystal structures tend to be more difficult to deform, so they are harder and have a high tensile strength.

Three types of pearlite are typically formed by heat treatment processes:

- **Furnace Cooling.** Coarse Pearlite (Pure Form: 15 Rockwell C or 240 BHN)

- **Air Cooling.** Medium Pearlite (Pure Form: 25 Rockwell C or 280 BHN)

- **Oil Quench.** Fine Pearlite (Pure Form: 38 Rockwell C or 380 BHN)

Unless a steel is eutectic composition, the pearlite is mixed with ferrite (Pure Form: 80 BHN) or cementite (Pure form: 1000 BHN). The mix between ferrite/cementite and pearlite in a plain carbon steel can provide an estimate of the hardness of the steel. The following formula can be applied to estimate the hardness of a hypoeutectoid plain carbon steel after heat treatment based on carbon content and type of cooling process:

$$H = H_F \left(\frac{0.77 - \%C}{0.77} \right) + H_P \left(\frac{\%C}{0.77} \right)$$

where H_F is the BHN for ferrite and H_P is the BHN for the particular type of pearlite that is formed.

The tensile strength of a plain carbon steel can be estimated by the formula:

$$UTS = 500 \cdot BHN \; psi$$

Water or brine can be used for cooling processes, but the resulting microstructure is quite different. *Martensite* (Pure Form: 55 Rockwell C or 700 BHN) results from austenite being cooled at a high rate. Martensite is extremely hard and brittle. The volume increases as much as 4% during the martensite transformation. These expansions, coupled with thermal gradients, result in internal stresses and can cause *quench cracking*.

The relative severity of quenching for different media can be approximated by:

relative severity	quenching media
5	agitated brine
1	still water
0.3	still oil
0.02	still air

In other words, agitated brine can cool an austenized sample 250 times faster than still air.

The different cooling rates and the resulting microstructure are best shown on a time-temperature-transformation (TTT) diagram as shown in Figure 10.8. If a steel sample is heated above the transformation temperature and cooled more quickly than the critical cooling rate, the resulting structure will contain martensite. If the sample is cooled more slowly, the resulting structure will contain pearlite.

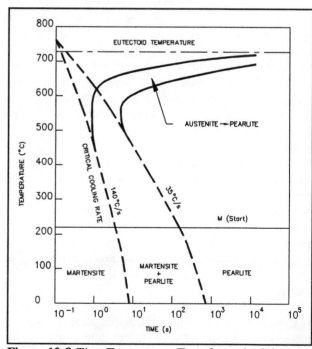

Figure 10.8 Time Temperature Transformation Diagram.

Example 10.4.1. What temperature is needed to completely austenize a sample of plain carbon steel with 0.4% carbon? Estimate the hardness after austenizing and oil quenching.
Solution. The sample is completely austenized at approximately 760°C. The hardness can be found using the equation for hypoeutectic steels:

$$H = 80\left(\frac{(0.77-0.4)}{0.77}\right) + 380\left(\frac{0.4}{0.77}\right) = 236 \; BHN$$

Example 10.4.2. Estimate the tensile strength of plain carbon steel with 0.2% carbon after austenizing and air cooling.
Solution. The tensile strength can be estimated based on the BHN. The BHN can be found as:

$$H = 80\left(\frac{(0.77-0.2)}{0.77}\right) + 280\left(\frac{0.2}{0.77}\right) = 132 \; BHN$$

The tensile strength is estimated by:

$$UTS = 500 \cdot BHN = 500\,(132) = 66000 \; psi$$

Heat treatment involves many different types of processes including annealing, stress relieving, normalizing, hardening, spheroidizing and case hardening.

- **Annealing.** A general term for restoration of the cold-worked or heat treated material to its original properties. It may be performed to increase ductility and reduce hardness and strength. It is also done to relieve residual stresses in a manufactured part to improved machinability. Annealing is done by heating the part to a temperature higher than the austenizing temperature. The material is "soaked" at that temperature for a period of time and allowed to slowly cool. The resulting structure is coarse pearlite (soft and ductile with uniform grains).

- **Normalizing.** To avoid excessive softening of steels, normalizing is performed. The part is heated to well above the austenizing temperature and is allowed to air cool. The resulting part will have a higher strength and hardness, but lower ductility than full annealing.

- **Spheroidizing.** A process used to improve the properties of high carbon steels. The process involves holding the part at a temperature very close to the lower critical temperature for an extended period (24 hours). These high temperatures permit the cementite to form small spheres that limit the

stress rising characteristics of the flakes of cementite. The result is a material with higher ductility and lower strength.

- **Stress Relieving.** Residual stresses in a part (induced by forming or machining) can be reduced by heating the material to a temperature close to but below the lower critical temperature and slowly air cooling it.

- **Tempering.** Previously hardened steels are heated to prescribed temperatures to reduce brittleness, increase ductility and toughness and reduce residual stresses. Temperatures below $400°F$ ($204.4\,°C$) are typically used for applications requiring high wear resistance and above $800°F$ ($427\,°C$) are used for applications requiring high toughness.

The typical heat treatment temperature ranges are illustrated in Figure 10.9.

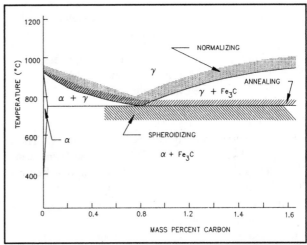

Figure 10.9 Typical Heat Treatment Temperatures for Steel.

Case hardening is applied in applications where thorough hardening is undesirable. Thorough hardening of parts is not desired in some applications such as gear teeth, cams, bearings, fasteners and dies since toughness is lost in hardening. A small surface crack could propagate rapidly through the part and cause total failure. A surface hardening process, case hardening charges the outer layer composition of an iron-based alloy and is followed by appropriate thermal treatment.

10.5 Alloy Steels
Various elements are added to steel to tailor its properties. Alloying elements can enhance the strength, hardness, toughness, wear resistance and many other properties. The most common alloying elements are presented in Table 10.1 with a summary of their effects.

Table 10.1 Steel Alloying Elements.

Element	Effect
Carbon	Improves hardenability, strength and wear resistance. Reduces ductility and weldability.
Chromium	Improves toughness, wear and corrosion resistance and high temperature strength.
Cobalt	Improves strength and hardness at elevated temperatures.
Lead	Improves machinability. Causes embrittlement.
Molybdenum	Improves hardenability, wear resistance, toughness, elevated temperature strength, and creep resistance.
Nickel	Improves strength, toughness and corrosion resistance.
Vanadium	Improves strength, toughness, abrasion resistance and hardness at elevated temperatures.

There are standard designations for steel alloys. The most widely used system is the AISI-SAE system (American Iron and Steel Institute and Society of Automotive Engineers). The first two digits indicate the alloying elements and the relative composition. The last two digits indicate the percent carbon by weight. For example, 1020 steel is a plain carbon steel with 0.2% carbon by weight. The characteristics of the AISI-SAE steel series is given in Table 10.2.

10.6 Stainless Steels

Stainless steels are used for applications where corrosion resistance, high strength, and heat resistance are important. These steels are characterized by a high chromium content. They are called *stainless* because a thin, air-tight film of chromium oxide forms on the surface to isolate the metal from corrosion. Chromium oxidizes very rapidly. The protective film reappears quickly if the surface is scratched. There are several types of stainless steels, identified by an AISI three-digit numbering system. The significance of the last two digits varies, but the first digit indicates the group. The numbering system is shown in Table 10.3

Table 10.2 Characteristics of AISI-SAE Steel Series.

AISI Number	Characteristics
10XX	Plain Carbon
13XX	Manganese - increases strength in as-rolled state and increases ductility after heat treatment
23XX-25XX	Nickel - Increase tensile strength without loss of ductility
3XXX	Nickel/Chromium - tough and ductile due to Nickel, wear and corrosion due to chromium
4XXX	Molybdenum - Significant increase in tensile strength and hardenability
5XXX	Chromium - high wear resistance
6XXX	Chromium/Vanadium - High yield strength, good fatigue properties
8XXX-9XXX	Chromium/Nickel/Molybdenum - Exhibits benefits of each

Table 10.3 Stainless Steels.

AISI Number	Characteristics
2XX	Chromium-Nickel-Manganese composition, non-hardenable, austenitic, non-magnetic
3XX	Chromium-Nickel composition, non-hardenable, austenitic, non-magnetic
4XX	Chromium composition, hardenable, martensitic and magnetic or non-hardenable, ferritic and magnetic
5XX	Chromium composition, heat resisting

The characteristics of the metallurgy of stainless steel are as follows:

• **Austenitic.** These stainless steels are nonmagnetic and do not harden by heat treatment. They are hardened by cold working. Austenitic stainless is the most ductile type of stainless steel. Typical applications are kitchen utensils, fittings and welded construction.

• **Ferritic.** This group is magnetic and has a superior corrosion resistance, but they are less ductile than austenitic stainless steels. They cannot be hardened by heat treatment. Typical applications are nonstructural applications in corrosive environments.

• **Martensitic.** The martensitic stainless steels do not contain nickel. These steels are hardenable by heat treatment. They are magnetic and exhibit high strength, hardness, fatigue resistance and ductility. However, they have only a moderate corrosion resistance. Typical applications are valves, springs and cutlery.

10.7 Cast Iron

Cast iron is basically an alloy of iron and carbon. It contains between 2% and 6.67% carbon. The high carbon content tends to make cast iron brittle. Cast iron cannot be rolled, drawn, or otherwise worked unless it is done at an elevated temperature. The desirable characteristics it incorporates are its low melting temperature and its castability (ease of pouring into complicated shapes.)

There are several basic types of cast iron. The characteristics are determined by carbon content and form, alloy content and heat treatment.

• **Gray Iron.** The carbon content is in a free state (flakes of graphite). This is the most widely used cast iron due to its machinability and high shear strength. The graphite is formed by the addition of silicon and phosphorus. The free graphite acts as a lubricant in the machining process.

• **White or Chilled Iron.** This product is made by casting gray iron against a metal heat sink. The localized rapid cooling results in a hard, abrasion-resistant surface with softer gray iron core. No free graphite is formed upon rapid cooling. Instead, the carbon forms cementite. Limited depths of chill are possible due to the physical limitations in transferring heat away from the molten metal.

• **Malleable Iron.** Malleable iron is made by heat treating white iron. Free graphite is formed by holding the white iron component at a high temperature. The resulting product has higher strength, ductility and machinability.

• Nodular Iron. Nodular iron or ductile iron contains carbon in the form of tiny nodules or spheres. The composition is similar to gray iron except the addition of

magnesium or cerium causes the formation of spheres of graphite rather than flakes. The nodular structure gives this material excellent impact properties.

10.8 Aluminum

Aluminum is an important engineering material because of its high strength to weight ratio, resistance to corrosion, high thermal and electrical conductivity, appearance, machinability, and formability. Aluminum is used for packaging, structures, electrical conductors (nearly all high-voltage transmission lines are aluminum) and consumer goods.

Table 10.4 Wrought Aluminum Alloy Designations.

Number	Alloy	Properties
1XXX	Commercially Pure	Corrosion resistant, high electrical and thermal conductivity, good workability, low strength
2XXX	Copper	High strength to weight ratio, low corrosion resistance
3XXX	Manganese	Good workability, moderate strength
4XXX	Silicon	Low melting point
5XXX	Magnesium	Good corrosion resistance, weldable, high strength
6XXX	Magnesium and Silicon	Good weldability, machinability and formability, corrosion resistant, medium strength
7XXX	Zinc	High strength

Various types of aluminum are used in the form of *wrought products* (made into various shapes by rolling, extrusion, drawing, and forging) and *cast alloys* (used for molding into various shapes). Wrought aluminum alloys are identified by four digits and a *temper designation* (indicating the processing of the material). The major alloying element determines the first digit. The various types of wrought aluminum are shown in Table 10.4.

The second digit refers to other alloying elements and the third and fourth digits indicate the amount of aluminum present in the alloy (e.g. 1070 contains 99.70% aluminum and 1090 contains 99.9% aluminum).

The designations for cast aluminum alloys also uses four digits. A decimal point is added between the third and fourth digit. In the 1XX.X series, the second and third digits indicate the aluminum content. The fourth digit, which is to the right of the decimal, indicates the product form: 1XX.0 indicates castings and 1XX.1 indicates ingot. In the other series, these digits vary in their usage. The cast aluminum alloy numbering system is shown in Table 10.5.

The temper designations used for aluminum apply to both wrought and cast alloys. The following letter designations are used:

F	As fabricated
O	Annealed
H	Strain hardened by cold work
T	Heat treated

Table 10.5 Aluminum Casting Alloys.

Number	Alloy	Properties
1XX.X	Commercially Pure	Corrosion resistant
2XX.X	Copper	High strength and ductility
3XX.X	Silicon	Good machinability (with copper or magnesium)
4XX.X	Silicon	Good castability, corrosion resistant
5XX.X	Magnesium	High strength
6XX.X	Unused	
7XX.X	Zinc	High strength, excellent machinability

Practice Problems

10.1   How are grain boundaries formed in a metal?

10.2   How does the addition of salt effect the solidus temperature of water?

10.3   What is the range of carbon content in typical steels?

10.4   What is the lowest possible austenizing temperature?

10.5   Why is oil sometimes used in quenching heat treated samples in preference over water?

10.6   Estimate the hardness of a specimen of plain carbon steel with 0.5% carbon (AISI 1050) after austenizing and subsequent oil quenching.

10.7   Estimate the tensile strength of a plain carbon steel with 0.3% carbon after austenizing and furnace cooling.

10.8   Recommend an annealing temperature for a steel with 0.2% carbon.

10.9   What metal can be alloyed with steel to improve its machinability?

10.10   What metal is alloyed with steel to improve its corrosion resistance?

10.11   What type of cast iron would be most likely used for a part requiring many difficult machining operations.

10.12   A C-clamp is to be made from cast iron and the manufacturer wants the part to resist being broken when dropped. What type of cast iron should be selected?

10.13   What type of wrought aluminum should be chosen for a part that is to be welded and exposed to salt spray (in a marine application)?

FLUID MECHANICS

Fluid Mechanics studies fluids at rest and in motion by applying principles of statics, kinematics and dynamics. This discipline covers a broad field including gases and liquids. Fluids are defined as substances that continuously deform when subjected to a shearing stress.

11.1 Fluid Properties
There are several common fluid properties that are used to describe fluid behavior.

• **Density** ρ is the ratio of mass m to volume V of a substance.

$$\rho = \frac{m}{V}$$

• **Specific Volume** υ is the volume occupied by a unit mass of substance.

$$\upsilon = \frac{1}{\rho}$$

• **Specific Weight** γ is the force of gravity on a mass per unit volume.

$$\gamma = \rho\, g$$

• **Specific Gravity** S is the ratio of the density of substance to the density of water.

$$S = \frac{\rho}{\rho_w}$$

• **Viscosity or Dynamic Viscosity** μ is one measure of the resistance to flow for a substance. Dynamic viscosity is the ratio of the shear stress to the velocity gradient (change in velocity over distance).

$$\mu = \frac{\tau}{du/dy} \cong \frac{\tau}{\Delta u/\Delta y}$$

• **Kinematic Viscosity** ν is another measure of resistance to flow. It is expressed as the ratio of dynamic viscosity to the density of a substance.

$$\nu = \frac{\mu}{\rho}$$

• **Bulk Modulus** K is a measure of the compressibility of a fluid.

$$K = -V\frac{\Delta p}{\Delta V}$$

Example 11.1.1. Two measurements of the velocity of water flowing in a pipe differ by 1.8 *m/s* over a 20 *mm* distance. What is the shear stress in the water? The dynamic viscosity of water under these conditions is 0.001 $N{\cdot}s/m^2$.

Solution. The definition of dynamic viscosity is rearranged algebraically to obtain:

$$\tau = \mu\frac{\Delta u}{\Delta y} = 0.001\ N{\cdot}s/m^2\ \frac{1.8\ m/s}{0.02\ m} = 0.09\ Pa$$

11.2 Fluid Statics
The study of fluid statics often involves the change of fluid pressure due to the change in depth in the fluid. Some applications for fluid statics include using manometers, finding the forces on submerged bodies and determining buoyancy.

The *pressure* associated with a fluid is the force that it exerts on a body per unit of area. There are several definitions of pressure that are depicted in Figure 11.1. Gages typically measure pressures relative to atmospheric pressure either as a *gage pressure* or as a *vacuum*

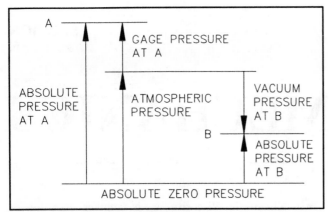

Figure 11.1 Pressure Definitions.

pressure. Typically, the relationship between absolute pressure and gage pressure is given by:

$$P_{abs} = P_{atm} + P_{gage}$$

Absolute pressures are measured relative to an absolute zero datum pressure.

If a fluid has a constant specific weight, pressure at any depth h in the fluid is given by:

$$p = \gamma h$$

where $p = 0$ at $h = 0$. This relationship is useful for measuring pressures by measuring the height of a column of fluid in a manometer.

Example 11.2.1. Use the depth measurements in the manometer to find the gage pressure of the water in the pipe shown in Figure 11.2. The density of water is 1000 kg/m^3. The specific gravity of mercury is 13.6.

Solution. Two points in the same fluid are located that are at the same relative depth. Equal pressures will be found at these points. In this case, A and B in the mercury will be used where $p_A = p_B$ or:

$$p_W + \rho_W g h_W = \rho_M g h_M + p_{ATM}$$
$$= S \rho_W g h_M + p_{ATM}$$

$$p_W + (1000)(9.8)(0.3) = (13.6)(1000)(9.8)(0.1) + p_{ATM}$$

$$P_{GAGE} = P_W - P_{ATM} = 10.4 \, kPa$$

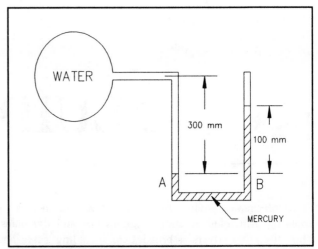

Figure 11.2 Manometer.

The principle of the variation of pressure with depth can be used to determine the force acting on a submerged object as shown in Figure 11.3. The forces acting on a

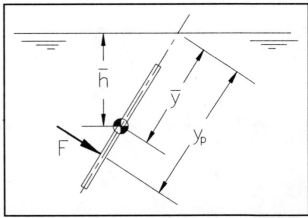

Figure 11.3 Static Force Acting on a Submerged Plate.

plane surface can be summed to give the resulting magnitude:

$$F = \gamma \bar{h} A$$

where F is the resultant force acting on one side of the object and $\bar{h}$ is the depth of the centroid of the surface relative to the surface and A is the surface area. The line of action of the force is at the center of pressure y_p given by:

$$y_p = \bar{y} + \frac{I}{\bar{y} A}$$

where $\bar{y}$ is the location of the centroid, A is the area of the submerged surface and I is the area moment of inertia of the area about its centroid.

Example 11.2.2. Find the force R that is necessary to hold the hold gate shut shown in Figure 11.4. The gate is 2 m wide (direction into the page).

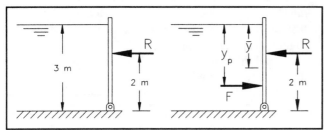

Figure 11.4 Submerged Gate.

Solution. The magnitude of force acting on the plate is given by:

$$F = \gamma g \bar{h} = 9.8(1000)(1.5) = 14700\ N$$

The center of pressure on the plate is given by:

$$y_p = \bar{y} + \frac{I}{\bar{y}A} = \bar{y} + \frac{\left(\dfrac{b h^3}{12}\right)}{\bar{y}A}$$

$$y_p = 1.5 + \frac{\left(\dfrac{(2)(3)^3}{12}\right)}{(1.5)(2)(3)} = 2.0\ m$$

Finally, the magnitude of the force R can be found by summing moments about the hinge:

$$\Sigma M_{HINGE} = 0$$

$$(R)(2)-(F)(1) = 2R - 14700 = 0$$

$$R = 7350\ N$$

Another issue in fluid statics is buoyancy. Floating objects will displace a volume of fluid that is equal to their weight. This is the principle that governs the floatation of boats and balloons. The weight of fluid W that is displaced by an object of volume V is given by:

$$W = \gamma V$$

where γ is the specific weight of the fluid.

Example 11.2.3. A barge has a rectangular bottom with dimensions 8 meters by 30 meters. It has a mass of 60,000 kg and is loaded with 200,000 kg of cargo. Find how far the barge will sink into the water.

Solution. The total weight of the loaded barge must equal the weight of the displaced water.

$$W = \gamma V$$

$$(60{,}000 + 200{,}000)(9.8) = (9.8)(1000)(8 \cdot 30 \cdot h)$$

$$h = 1.1\ m$$

11.3 Fluid Dynamics

The study of fluid dynamics considers the *flow* of fluids. The main issues in fluid flow center on the velocity, pressures and forces necessary to cause fluid to move. Most fluid flow problems are analyzed in the form of an imaginary system into which and from which the fluid flows often called a *control volume*. There are three main principles that govern fluid dynamics: *conservation of mass, conservation of momentum* (Newton's Second Law) and *conservation of energy*. It must be emphasized that the following development only applies to fluids that can be regarded as incompressible. Water and oil are examples of fluids typically regarded as incompressible.

Conservation of mass is described by the *continuity equation*:

$$A_1 v_1 = A_2 v_2$$

where A is the area that the fluid flows through, v is the velocity of the fluid and the subscripts refer to the point where the fluid enters and exits the system.

Conservation of momentum is described by the *momentum equation*:

$$\Sigma F = \rho Q(v_2 - v_1)$$

where ΣF is the summation of forces acting on the system, ρ is the density of the fluid and Q is the volumetric flowrate ($Q = A v$). Note that the forces and velocities are vector-valued quantities.

Conservation of energy is described by the *energy equation*, also known as the *Bernoulli equation*:

$$\frac{v_2^2}{2g} + \frac{P_2}{\gamma} + z_2 = \frac{v_1^2}{2g} + \frac{P_1}{\gamma} + z_1$$

where p is the pressure of the fluid and z is the elevation of the system relative to a datum. It will be assumed that flow is *steady state and incompressible* with a *uniform velocity profile*.

Example 11.3.1. Water flows through a 100 mm diameter pipe at 8 m/s. Downstream, the pipe is reduced in diameter to 40 mm. Find the velocity of the water in the smaller diameter.

Solution. The continuity equation is applied directly:

$$A_1 v_1 = A_2 v_2$$

$$\frac{\pi d_1^2}{4} v_1 = \frac{\pi d_2^2}{4} v_2$$

$$\therefore v_2 = v_1 \frac{d_1^2}{d_2^2} = 8 \cdot \frac{100^2}{40^2} = 50 \text{ m/s}$$

Example 11.3.2. A stream of water moving at 100 meters/second flows from a 10-mm-diameter pipe and is directed at a vertical plate, as shown in Figure 11.5. What force must be applied to the back of the plate to hold it in the position shown?

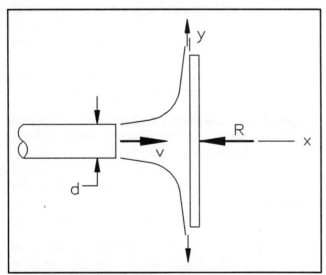

Figure 11.5 Stream of Water on a Plate.

Solution. The reaction on the plate can be found by applying the momentum equation. It should be noted that the velocity of the water in the x direction at the plate is zero. Since the force and velocity terms are vector-valued, the momentum equation is applied in the x direction to yield:

$$-R_x = \rho Q (v_{2x} - v_{1x})$$

$$R_x = \rho (A_1 v_{1x}) v_{1x}$$

$$R_x = \rho A_1 v_{1x}^2 = 1000(\pi \cdot 0.005^2)100^2 = 785 \ N$$

Example 11.3.3. One of the most important applications for a manometer is measuring fluid velocities with a *pitot tube*. The pitot tube measures fluid velocity indirectly by the height of a manometer column as shown in Figure 11.6. The manometer fluid is mercury, $S = 13.6$, and the height of the column h is 150 mm. Find the velocity of the water in the pipe.

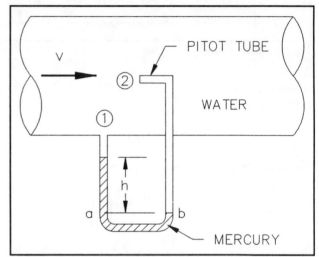

Figure 11.6 Use of a Pitot Tube for Measuring Fluid Velocity.

Solution. The energy equation is applied with the two reference points at the entrance to the pitot tube and at the wall of the pipe.

$$\frac{v_2^2}{2g} + \frac{p_2}{\gamma} + z_2 = \frac{v_1^2}{2g} + \frac{p_1}{\gamma} + z_1$$

The velocity of the water is zero at the entrance to the pitot tube and the difference in elevation is negligible, so the energy equation can be simplified to:

$$\frac{v_1^2}{2g} + \frac{p_1}{\gamma} = \frac{p_2}{\gamma}$$

The velocity of the water can be found by:

$$v_1 = \sqrt{\frac{2g}{\gamma}(p_2 - p_1)}$$

The difference in pressures can be found by observing the pressures are equal at points a and b in the manometer:

$$p_a = p_b$$

$$p_1 + \gamma_m h - p_2 + \gamma_w h$$

Making substitutions for the specific weights of mercury and water:

$$p_2 - p_1 = h(13.6 - 1)(9800)$$

$$= 0.150(12.6)(9800)$$

$$= 18.5 \, kPa$$

Finally, the velocity of the water is given by:

$$v_1 = \sqrt{\frac{(2)(9.8)}{1000}(18.5 \times 10^3)} = 19 \, m/s$$

Practice Problems

11.1 What are the units of kinematic viscosity in the metric system?

11.2 A cylindrical chamber is 12 inches long and 2 inches in diameter. The chamber is completely filled with a fluid, closed at one end, and has a piston at the other end. A 1200-pound compressive force on the piston causes it to move by 0.010 inch. Find the bulk modulus of the fluids.

11.3 Atmospheric pressure at sea level is approximately 14.7 psi. A vacuum pump is rated at drawing a 10-psi vacuum. What is the absolute pressure created by the vacuum pump?

11.4 The manometer shown in Figure 11.2 is used in another application where the height of the water column is 250 mm and the height of the mercury column is 50 mm. Find the pressure of the water.

11.5 A closed steel box with dimensions 4 feet by 4 feet by 2 feet is submerged in water with the largest surface facing up. The top surface of the box is 20 feet under water. Find the force acting on a vertical face of the box.

11.6 A canoe weighs 60 pounds. A 130-pound woman gets into the canoe while it is floating in the water. What volume of water does the canoe displace?

11.7 A pipe carries water from a large open tank elevated 50 meters above the pipe opening. Find the velocity of the water exiting the pipe.

11.8 What water pressure is needed to pump water up to the top of a 300-foot-tall building?

11.9 A stream of water is directed at a plate as shown in Figure 11.5. Find the pipe diameter needed to produce a 1000 N force on a plate with water moving at 60 m/s.

11.10 A mercury-filled pitot tube is used to measure the velocity of water in a pipe. A 10-mm mercury column is observed. What is the velocity of the water?

Part 4

Product Design

ENGINEERING DRAWING

Engineering drawings are the graphical representations of ideas. Graphical representations are recognized as an efficient and nearly universal means of communicating designs, instructions and plans. These documents graphically describe shapes, sizes and manufacturing processes used in a product. Drawings are the primary method used to control production. Standard practices are used in engineering drawings to avoid confusion and improve the effectiveness of communication. This chapter provides an overview of some of the recommended practices and standards used in the graphical language of engineering drawings.

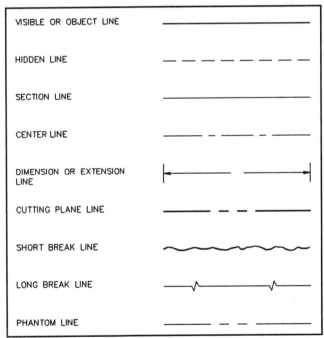

Figure 12.1   Types of Lines Used in Engineering Drawings.

12.1 General Drawing Standards

Engineering drawings make use of standard lines to aid in showing the details associated with a part. The standard lines are shown in Figure 12.1 and their application is described in Table 12.1.

Table 12.1 Application of Various Line Types.

Line Type	Application
Object or Visible	Visible edges of parts
Hidden	Hidden edges of parts that are not directly visible in a view
Center	Center positions of holes, shafts, radii and arcs
Phantom	Position and relationship of adjacent parts and alternate positions of moving parts
Dimension and extension	Size and location of part features
Leader	Special details, notes or specifications
Cutting Plane	Position and path of an imaginary cut made to form a sectional view
Short Break	End of the partially illustrated portion of a small detail
Long Break	End of the partially illustrated portion of a large detail

There are two systems of projection recognized internationally for multiview drawings. The American system places the front view under the top view. The European system places the top view under the front view. The American system is known as *third angle projection* and the European system is known as *first angle projection*. The terms are derived from the trigonometric quadrants relative and the front and top viewing planes of a part as illustrated in Figure 12.3. Figure 12.2 shows the standard symbols used on drawings to indicate the projection convention in use.

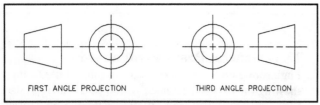

Figure 12.2 Drawing Symbols for Type of Projection.

There are six *principal views* for a drawing in third angle projection. The standard arrangement for these views is shown in Figure 12.4. A minimum number of views should be used to represent any part in an engineering drawing. All views are assumed to be rotated by **90°** from one another unless otherwise indicated on the drawing. The view that shows the most detail is typically selected as the front view. The front view is shown with an accompanying top or right side view to show detail that cannot be seen in the front view. If three views of a part are needed, the most common arrangement is to show the top, front and right side views in the standard arrangement shown in Figure 12.4. No superfluous views should be included in a drawing.

Auxiliary views are used to show the true size and shape of features not parallel to any principal views. Primary auxiliary views are projected from a principal view. Secondary auxiliary views can also be used by making a

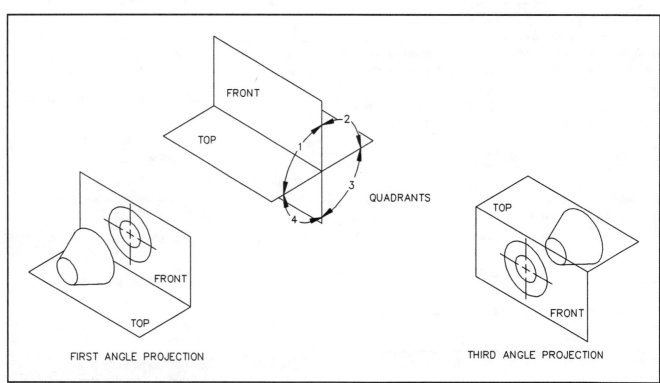

Figure 12.3 Standards for Multiview Projection.

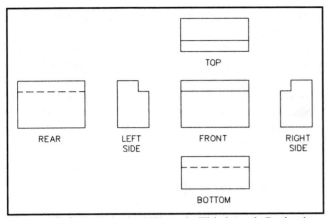

Figure 12.4 Six Principal Views in Third Angle Projection.

projection from a primary auxiliary view. Figure 12.5 shows an example of a slot seen in its true size and shape (TSS) in a primary auxiliary view.

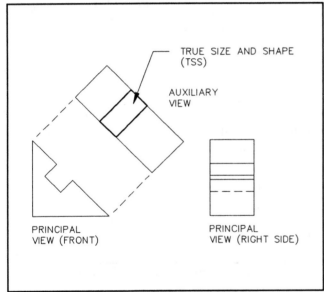

Figure 12.5 Auxiliary View.

Sectional views are drawn to show interior details. Sectional views are often clearer than exterior views which contain numerous hidden lines. The location and position of a sectional view is indicated by a cutting plane line and arrows indicating the line of sight in the section. Section lines are used to show the solid material in a sectional view. General sectioning is indicated by thin lines at a **45°** angle. Special sectioning symbols may also be used for specific materials. An example of a sectional view is shown in Figure 12.6.

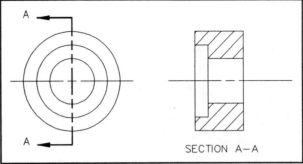

Figure 12.6 Sectional View.

More than one line may occupy the same position in a view as illustrated in Figure 12.7. The following *precedence of lines* should be applied in these cases:

a. Object lines take precedence over hidden lines and center lines.

b. Hidden lines take precedence over center lines.

c. Cutting plane lines take precedence over centerlines when showing the path of a sectional view.

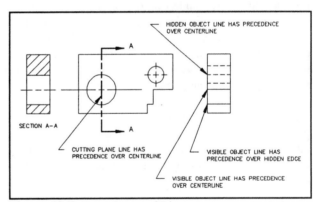

Figure 12.7 Precedence of Lines.

12.2 Dimensioning

Dimensions describe the details of a part so it can be constructed to the proper size. They must show the sizes and locations necessary to manufacture and inspect the part. Dimensions are placed between points that have a specific relationship to each other to ensure the function of the part. More than one view may illustrate a feature. The dimension of the feature should be placed in the view that best describes the feature. Dimensions are not repeated in different views.

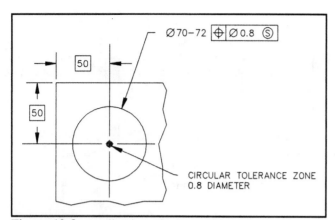

Figure 12.8 Application of a Basic Dimension.

All dimensions are subjected to a tolerance (amount of permissible variability unless noted as *basic dimensions* or *reference dimensions)*. Basic dimensions are identified by an enclosing frame symbol. The *basic dimension* is the theoretically exact size or location of a feature. It is the basis from which permissible variations are established by tolerances. Figure 12.8 shows a hole located with basic dimensions. A geometric tolerancing symbol is applied to the diameter dimension indicating that the true center of the circle lies in a circular tolerance zone. Reference dimensions are supplied for information only and may be used for checking purposes. They represent intended sizes, but they do not govern the manufacture of the part. Figure 12.9 shows an example of a part with three holes dimensioned in a series. The dimension that is not to be used in manufacturing the part is marked REF (for reference purposes only). Tolerances are applied to the 20 mm and 60 mm dimensions. These two dimensions are used to control the manufacturing process applied to the production of the part.

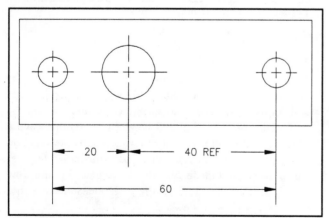

Figure 12.9 Application of a Reference Dimension.

Dimensions are related in three primary methods. *Chain dimensions* are used when the tolerance between adjacent features is more important than the overall tolerance accumulation. *Datum dimensions* are used when the location of features must be controlled from a common reference plane. *Direct dimensioning* is applied to control specific feature locations. These three types of dimensioning methods are shown in Figure 12.10. In this case, a general tolerance is applied to all dimensions.

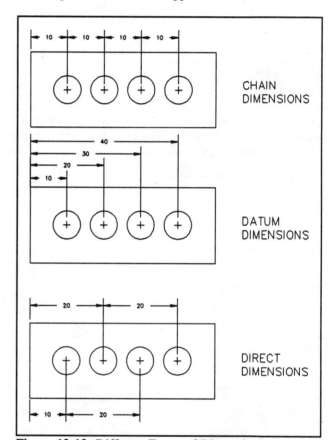

Figure 12.10 Different Types of Dimensioning.

12.3 Tolerancing

The methods for determining tolerances in manufactured parts are detailed in other sections. The methods for representing various tolerances are presented here. The major terms used in tolerancing are defined as follows:

• **Tolerance** is the permissible variation in a dimension. It is the difference between the largest and smallest acceptable sizes for a feature.

• **Limits** are the extreme allowable sizes for a feature.

- **Allowance** is the minimum clearance between mating parts. In the case of a shaft and mating hole, it is the difference in the diameters of the largest shaft and smallest hole.

- **Nominal Size** is an approximate size used for naming purposes. A "9/32" drill is an example of a nominal size.

- **Basic Size** is the exact theoretical size from which the limits are derived. The basic size for a 9/32 drill is 0.28128.

12.4 Fits

A *fit* signifies the type of clearance that exists between mating parts. There are four types of fit: *clearance, interference, transition* and *line*. Clearance fits provide some gap between mating parts. Interference fits have no clearance, force is required for the assembly to occur. Transition fits are toleranced to result in either a clearance or an interference. Line fits result in contact or clearance upon assembly.

Standard systems of fits are applied to holes and shafts that govern the tolerances according to the basic size of the components. Fits can either be based on a *standard hole system* or a *standard shaft system*. In a standard hole system, the smallest allowable hole is taken as the basic size from which the limits of tolerance are applied. In a standard shaft system, the largest allowable shaft is taken as the basic size.

There are standardized American National Standard and metric sizes for holes and shafts to achieve different types of fits. The American Standard system is a set of *classes* of fits based on the basic hole system. The types of fit covered by this standard are:

> RC - Running and Sliding Fits
> LC - Clearance Locational Fits
> LT - Transition Locational Fits
> LN - Interference Locational Fits
> FN - Force and Shrink Fits

Tables of standard sizes and tolerances are needed to use this system. Since the system is organized on a hole basis, the basic shaft size and the type of fit is needed to determine the dimension and tolerance for the mating parts (e.g., an RC 4 fit refers to a *close running fit* whereas RC 9 refers to a *loose running fit*. Tables will supply the tolerances (in number of thousandths) to add to the basic size to determine

the upper limit on the hole and to subtract from the basic size to determine the upper and lower limits on the shaft. The standard tables are too extensive to be reprinted here, but should be available in any text on engineering drawing.

Example 12.4.1. Find the dimensions and tolerances for the 2.500 hole with an RC 9 fit.
Solution. A table of American National Standard fits provides the following:

LIMITS OF CLEARANCE	HOLE	SHAFT
9.0	7.0	-9.0
20.5	0	-13.5

The lower limit of the hole has no tolerance since it represents the basic size. The limit dimensions on the hole are found as:

UPPER LIMIT: 2.5000 + 0.0070 = 2.5070
LOWER LIMIT: 2.5000 + 0.0000 = 2.5000

The limit dimensions on the shaft are found as:

UPPER LIMIT: 2.5000 - 0.0090 = 2.4910
LOWER LIMIT: 2.5000 - 0.0135 = 2.4865

The limits of clearance represent the largest and smallest clearances that can result in assembly. They can be readily verified by taking the extreme shaft and hole combinations:

LARGEST
CLEARANCE: 2.5070 - 2.4865 = 0.0205

SMALLEST
CLEARANCE: 2.5000 - 2.4910 = 0.0090

The metric system of fits is organized in a fashion similar to the American National Standard. The system can operate on a *hole basis* or *shaft basis*. An International Tolerance (IT) Grade is associated with a particular size and the level of accuracy. For example, the designation 40H8 refers to a 40-mm basic size hole with an IT grade of 8. The designation 40f7 refers to a 40-mm basic size shaft with an IT grade of 7. The combination of 40H8/f7 refers to a particular fit, in this case a close-running fit. Tables of standard values are needed to determine the appropriate tolerances.

12.5 Geometric Tolerances

Geometric tolerances state the maximum allowable deviation of a form or a position from the perfect geometry implied by a drawing. These tolerances specify the diameter or the width of a tolerance zone necessary for a part to meet its required accuracy. Figure 12.11 shows the various symbols used to specify the required goemetric characteristics of dimensioned drawings. A modifier is used to specify the limits of size of a part when applying geometric tolerances. The *maximum material condition* (MMC) indicates that a part is made with the largest amount of material allowable (i.e. a hole at its smallest permitted diameter or a shaft at its largest permitted diameter). The least material condition (LMC) is the converse to the maximum material condition. *Regardless of feature size* (RFS) indicates that tolerances apply to a geometric feature for any size it may be. Many geometric tolerances or *feature control symbols* are stated with respect to a particular datum or reference surface. Up to three datum surfaces can be given to specify a tolerance. Datum surfaces are generally designated by a letter symbol. Figure 12.12 gives examples of the symbols used for datum planes and feature control symbols.

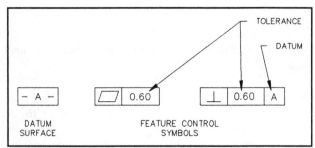

Figure 12.12 Feature Control Symbols.

	TOLERANCE	CHARACTERISTIC	SYMBOL
INDIVIDUAL FEATURES	FORM	STRAIGHTNESS	—
		FLATNESS	▱
		CIRCULARITY	○
		CYLINDRICITY	⌭
INDIVIDUAL OR RELATED FEATURES	PROFILE	LINE	⌒
		SURFACE	⌓
RELATED FEATURES	ORIENTATION	ANGULARITY	∠
		PERPENDICULARITY	⊥
		PARALLELISM	//
	LOCATION	POSITION	⊕
		CONCENTRICITY	◎
	RUNOUT	CIRCULAR RUNOUT	↗
		TOTAL RUNOUT	⌰
NOTES	⌀ DIA	Ⓜ MMC	Ⓛ LMC	Ⓢ RFS

Figure 12.11 Geometric Tolerancing Symbols.

12.6 Tolerances of Location

Tolerances of location are concerned with specifying requirements for position, concentricity and symmetry. Toleranced location dimensions are often used to specify the acceptable variability in a position. Figure 12.13 shows the center of a hole located with toleranced location dimensions. An alternative way of representing the location of the center is *true-position dimensioning*. Basic dimensions (the boxed quantities) give the exact untoleranced dimensions used to locate true positions. A circular tolerance zone is implied by the feature control symbol which indicates that the true position of the hole can vary within a 0.8-mm-diameter tolerance zone, regardless of the size of the hole. In both cases, the diameter of the hole may vary within the range of 112 mm to 113 mm. Toleranced location dimensions permit larger variation from the basic dimensions at the corners of the square tolerance zone. The corners of the square are permitted to vary up to $0.4\sqrt{2}$ mm or 0.48 mm from the basic center location. This situation is not necessarily a problem as long as this is the intent of the designer.

Concentricity is a feature that specifies the relationship between one cylinder and another. The specification of concentricity is typically the permitted variation between centerlines of two cylinders. One cylinder is flagged as the datum (indicating that its centerline will be used as a reference) and the permissible variation of the centerline of the other cylinder is specified. Figure 12.14 shows a larger diameter cylinder flagged as a datum. The centerline of the smaller diameter cylinder is permitted to vary within a cylindrical tolerance zone with a diameter of 0.3 mm.

Symmetry is also a location tolerance. A feature is symmetrical when it has the same contour and size on opposite sides of a central plane. A symmetry tolerance is always specified with respect to a datum plane. Figure 12.15 shows a slot which must lie between two parallel planes 0.8 mm apart, regardless of the size of the slot. The planes are equally spaced with respect to the center plane of datum B, regardless of the size of width of the part. The center plane of datum B is also treated as perpendicular to datum plane A.

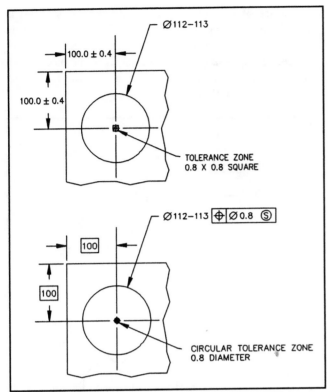

Figure 12.13 Position Tolerancing.

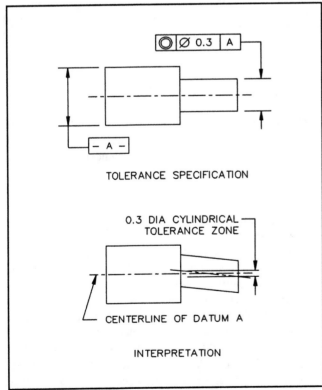

Figure 12.14 Concentricity Tolerancing.

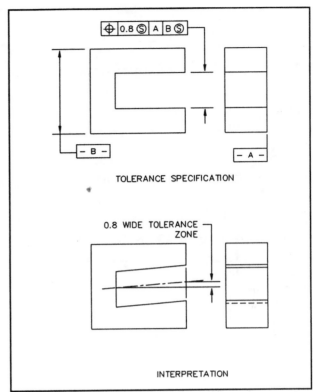

Figure 12.15 Symmetry Tolerancing.

12.7 Tolerances of Form

There are four tolerances that specify the permitted variability of forms: *flatness, straightness, roundness* and *cylindricity*. Form tolerances describe how an actual feature may vary from a geometrically ideal feature.

A surface is ideally flat if all its elements are coplanar. The *flatness* specification describes the tolerance zone formed by two parallel planes that contain all the elements on a surface. A 0.5-mm tolerance zone is described by the feature control symbol in Figure 12.16. The distance between the highest point on the surface to the lowest point on the surface may not be greater than 0.5 mm.

A surface is perfectly straight if all of its elements are colinear. *Straightness* is specified by two parallel lines that contain all the elements of a surface. A straightness tolerance is typically applied to cylindrical features. A 0.01-mm tolerance zone is described by the feature control symbol in Figure 12.17. All elements on the surface must lie between two parallel lines spaced 0.1 mm apart where the two lines and the axis of the cylinder share a common plane.

109

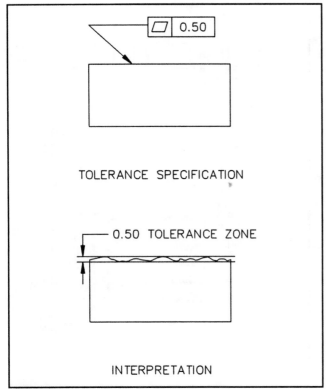

Figure 12.16 Flatness Tolerancing.

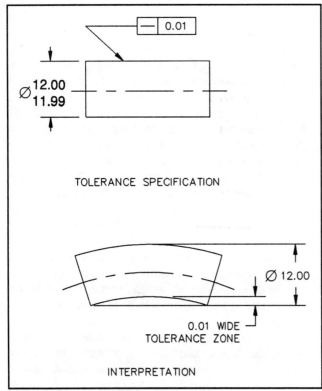

Figure 12.17 Straightness Tolerancing.

A surface of revolution (such as a cylinder or cone) is round when any cross section normal to its axis rotation is a circle. A *roundness* specification describes a ring-shaped tolerance zone that contains all of the elements in the cross section. Figure 12.18 illustrates a 0.1-mm cylindricity specification.

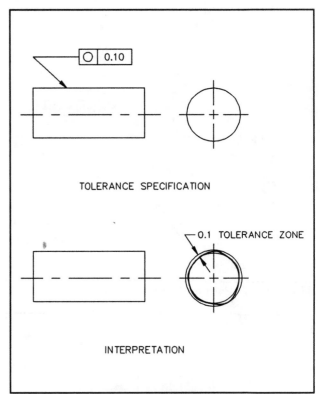

Figure 12.18 Roundness Tolerancing.

Cylindricity is a combination of roundness and straightness. A cylindricity specification is the tolerance permitted on the radius of a cylinder as shown in Figure 12.19.

12.8 Tolerances of Orientation

Parallelism and perpendicularity are examples of tolerances of orientation. A *parallelism* tolerance specifies a tolerance zone defined by two parallel planes that are mutually parallel to a datum. All elements of the toleranced surface must lie within the specified tolerance zone. A *perpendicularity* tolerance specifies a tolerance zone defined by two parallel planes that are mutually perpendicular to a datum. All elements of the toleranced surface must lie within the specified tolerance zone. An example of a perpendicularity specification is shown in Figure 12.20.

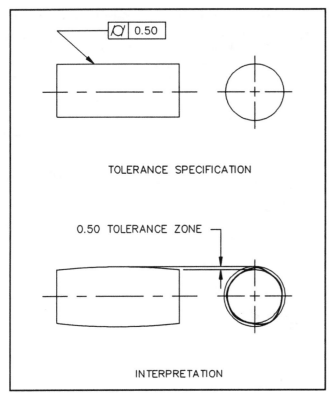

Figure 12.19 Cylindricity Specification.

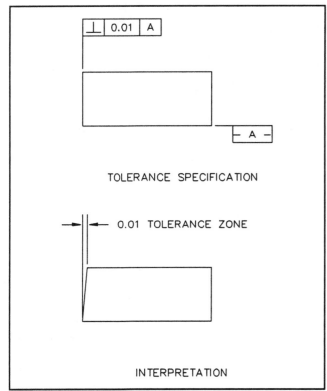

Figure 12.20 Perpendicularity Specification.

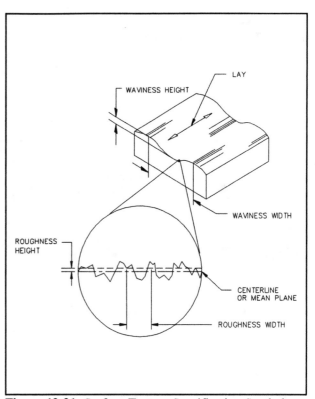

Figure 12.21 Surface Texture Specification Symbol.

12.9 Surface Texture

Surface texture is the variation of height, width and orientation of the irregularities on a surface. Surface texture can strongly affect the performance of a part in service. Surface texture specifications are critical to assuring the proper function of parts such as bearings or dies. Figure 12.21 illustrates a standard surface texture specification symbol. The important terms used in surface texture specification are defined below:

• **Roughness** refers to the finest irregularities in a surface. Roughness is strongly dependent on the type of manufacturing process used to generate a surface.

• **Roughness Height** is the arithmetic average deviation from the mean plane or centerline of a surface. It is typically measured in microinches or micrometers.

• **Roughness Width** is the spacing between the successive peaks and valleys that form the roughness of a surface.

• **Cutoff** is the sampling length used for calculation of the roughness height. When it is not specified, a value of 0.8 mm (0.030 inch) is assumed.

• **Waviness** is the widely spaced, repeated variation on a surface. The *Waviness Width* must exceed the cutoff length. The *Waviness Height* is the peak-to-valley distance between waves.

• **Lay** is the direction of the surface pattern. This is dependent on the method used to generate the surface. Figure 12.22 illustrates the standard lay designations.

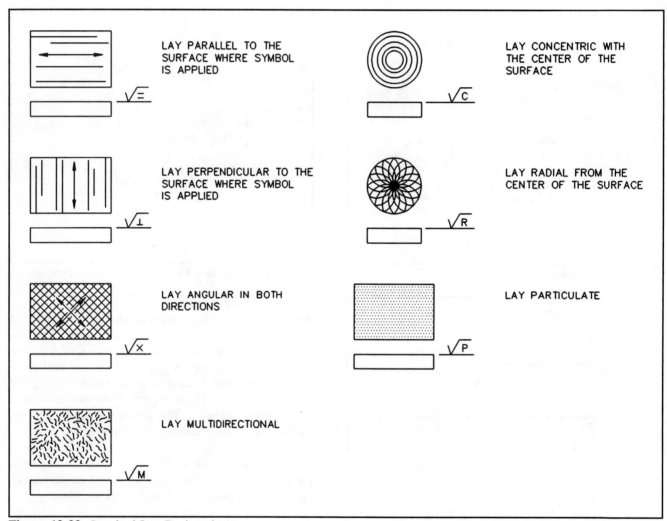

Figure 12.22 Standard Lay Designations.

Practice Problems

12.1 Draw the six principal views in first angle projection of the object shown in Figure 12.4.

12.2 Use the views in Figure P12.1 to create the auxiliary view of the object that shows the hole in its true size and shape.

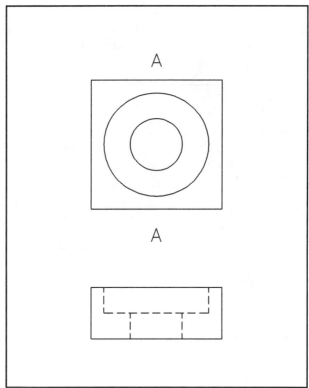

Figure P12.2 Problem 12.3.

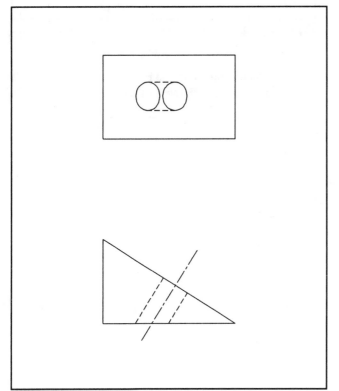

Figure P12.1 Problem 12.2.

12.3 Draw the sectional view passing through points A-A in Figure P12.2.

12.4 What type of dimension has no tolerance applied to it?

12.5 The part in Figure P12.3 must be manufactured so that the location of the top surface of each step must be accurately located with respect to the edge 1-2. What type of dimensioning should be used?

12.6 A 1-inch diameter pin is placed in a 1.3-inch-diameter hole. The dimensions of both the pin and the hole may vary ±0.1 inches. What is the allowance?

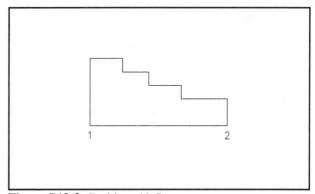

Figure P12.3 Problem 12.5.

12.7 Is the American National Standard based on shafts, holes or both?

12.8 Is the International Tolerance Grade Based on shafts, holes or both?

12.9 Interpret the symbols in the following IT designation: 50f6.

113

12.10 What are the geometric tolerancing symbols for individual features?

12.11 Give the geometric dimensioning and tolerancing specification for the situation shown in Figure P12.4.

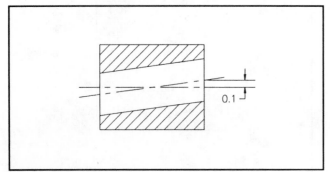

Figure P12.4 Problem 12.11.

12.12 Give the geometric dimensioning and tolerancing specification for the situation shown in Figure P12.5.

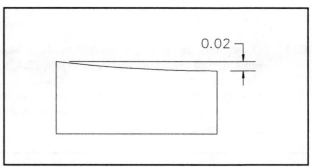

Figure P12.5 Problem 12.12.

12.13 Identify the indicated parts of the surface texture specification shown in Figure P12.6.

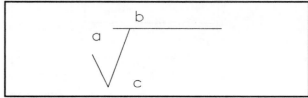

Figure P12.6 Problem 12.13.

114

TOLERANCES

Modern mass production calls for parts made at remote locations to be brought together for assembly and to fit properly without modification. Manufacturers depend on any part being able to be assembled with the intended mating part. Assembly would be no problem if all parts could be made exactly to size. Some parts, such as gage blocks, can be made to be very close to a target dimension, but such accuracy is very expensive.

Exact sizes of parts are impossible to produce. For practical reasons, parts are made to varying degrees of accuracy depending on their functional requirements. A *tolerance* refers to the degree of accuracy that is required in a dimension. In general, the cost of manufacturing a component increases with smaller tolerances on its dimensions.

13.1 Terminology of Tolerancing

Figure 13.1 shows an assembly of two mating parts. Tolerances are assigned to the parts to control the dimensions so that a random selection of any shaft will fit in any hole. In Figure 13.1, the maximum diameter shaft is shown solid and the minimum diameter shaft is shown in dashed lines. The loosest fit, when the smallest shaft and largest hole are assembled, is shown in Figure 13.2. The condition for the tightest fit is shown in Figure 13.3. The following terms are commonly used in tolerancing :

• **Nominal Size** - The dimension used for general identification of a part. In Figure 13.1, the nominal size of both the shaft and the hole is 1.250.

• **Limits** - The maximum and minimum sizes of the toleranced dimension. In Figure 13.1 the limits for the shaft are 1.247 and 1.248.

• **Allowance** - The minimum clearance between mating parts in the maximum material condition (MMC) is the *allowance*. In Figure 13.3, the allowance is the difference

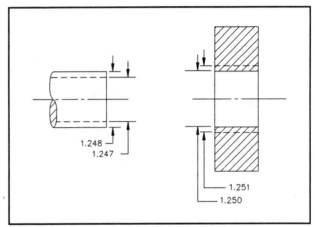

Figure 13.1 Tolerancing of Mating Parts.

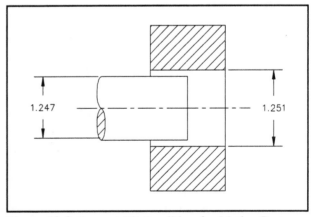

Figure 13.2 Loosest Fit Condition for Mating Parts.

between the diameters of the largest shaft and smallest hole.

• **Tolerance** - The total amount by which a dimension may vary or the difference between the limits is the *tolerance*. The difference in diameters, 0.001, is the tolerance on the diameter of the shaft shown in Figure 13.1.

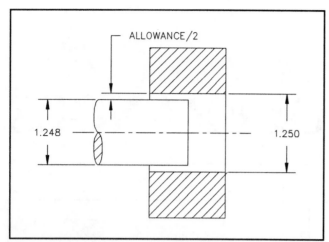

Figure 13.3 Tightest Fit in Mating Parts.

13.2 Specification of Tolerances

Figure 13.4 shows three methods that can be used to express tolerances in dimensions: limit dimensioning, unilateral tolerances and bilateral tolerances.

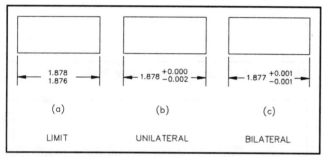

Figure 13.4 Different Methods for Expressing Tolerances.

• **Limit Dimensioning** - The maximum and minimum sizes of a feature are specified as shown in Figure 13.4(a). The maximum value is placed over the minimum value.

• **Unilateral Tolerances** - A nominal size is followed by a plus and minus expression of tolerance where variations are allowed in only one direction from the nominal size. Figure 13.4(b) shows an example of unilateral tolerances where all the tolerance is toward the smaller size.

• **Bilateral Tolerances** - A nominal size is followed by a plus and minus expression where variations are permitted in both directions from the nominal size. Figure 13.4(c) illustrates a dimension with bilateral tolerances.

Two criteria must be used for determining tolerance for a dimension:

• The tolerance should be chosen to permit randomly selected mating parts to assemble.

• The tolerance should be as large as possible. The cost of producing any manufactured component increases with smaller tolerances.

13.3 Tolerances for 100% Interchangeability

The most common method of determining the tolerance on a dimension calls for parts to be 100% interchangeable. Any random combination of mating parts will be guaranteed to assemble. The extreme or most difficult conditions for assembly are used to find the unknown tolerance. In the extreme condition, internal dimensions are taken at the minimum value and external dimensions are taken at the maximum value. A *path equation* is used to add signed dimensions to find the value of an unknown tolerance.

Example 13.3.1. Figure 13.5 shows a car radio tuner knob (k) being assembled with a bearing (b) and a spacer (s) in a cavity (c). In order for the knob to turn freely, a 0.003 clearance (g) must exist between the knob flange (a) and the top of cavity. Find the tolerance X of the depth of the cavity.

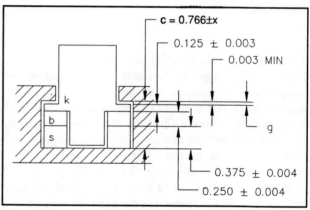

Figure 13.5 Car Tuner Knob Assembly.

Solution. A sign convention will be used where dimensions as positive going from the bottom to the top. Starting at the bottom of the cavity, the signed dimensions for the extreme conditions are added to form a closed path. The extreme conditions are found when the maximum size knob, bearing and spacer are combined with a minimum clearance (*g*) and minimum cavity (*c*):

$$k_{max} + b_{max} + s_{max} + g_{min} - c_{min} - 0$$

or:

$$0.128 + 0.379 + 0.254 + 0.003 - 0.766 - X = 0$$

$$X = 0.002$$

13.4 Statistical Interchangeability

Another approach to determining the tolerances on a dimension is to assure that a large percentage of parts are to be interchangeable. Statistical tolerancing methods result in larger tolerances at the expense of having a small percentage of mating parts that cannot be assembled. Statistical tolerancing can be applied to assemblies where the following assumptions are valid:

• All dimensions in the assembly are normally distributed.

• All dimensions are generated independently of each other. No manufacturing process used to create a feature on a part can depend on the size of a feature in a mating part.

• A random selection of parts is used in each assembly.

• The nominal size of feature is equal to the mean or average size of the feature.

In most mass production operations these assumptions can be taken as valid.

The statistical tolerance technique is applied by letting bilaterally toleranced dimension to be equal to the mean dimension, $\overline{X}$, plus or minus three standard deviations of the dimension, 3σ. If a nominal dimension was 1.000 and the bilateral tolerance was 0.003, statistical tolerancing would assume:

$$1.000 \pm 0.003 = \overline{X} \pm 3\sigma$$

where 0.001 would be considered the standard deviation. Furthermore, if T is the total tolerance spread in a bilaterally toleranced dimension, then the standard deviation can be considered to be given by:

$$\sigma = \frac{T}{6}$$

If the process generating these components generates dimensions that are normally distributed, then standard statistical analysis suggests that the mean dimension plus or minus three standard deviations represents the range of 99.73% of all possible dimensions. In other words this range describes almost all of the parts created by the process.

The statistical range of possible dimensions can be used to find appropriate tolerances for parts in an assembly. Two theorems from elementary statistics must be applied:

Theorem 1: The average dimension of an assembly is determined by the sum of the dimension of the components used in the assembly.

Theorem 2: The variance (the square of the standard deviation, σ^2) of the dimension of an assembly is determined by the sum of the variances of the dimensions of the components used in the assembly.

Example 13.4.1. Reconsider Example 13.3.1 using statistical tolerancing techniques. Find the unknown tolerance, X, in Figure 13.5 by applying the assumptions of statistical tolerancing.

Solution. The solution focuses on the gap, g, as being the variable dimension in the assembly. If the nominal dimensions are assumed to represent average dimension, then the average dimension of the gap can be found by application of Theorem 1:

$$\overline{k} + \overline{b} + \overline{s} + \overline{g} - \overline{c} = 0$$

or

$$0.125 + 0.375 + 0.250 + \overline{g} - 0.766 = 0$$

$$\overline{g} = 0.016$$

Since the minimum gap in the assembly is given as 0.003 and all of the tolerances are bilateral, the gap dimension can be taken to vary bilaterally from the average gap in a bilateral fashion:

$$g = 0.016 \pm 0.013$$

This suggests that the overall tolerance spread of the gap can be found as:

$$T_g = 0.026$$

The overall tolerance of the gap is determined by the tolerances of the components in the assembly as given by Theorem 2:

$$\left(\frac{T_g}{6}\right)^2 = \left(\frac{T_k}{6}\right)^2 + \left(\frac{T_b}{6}\right)^2 + \left(\frac{T_s}{6}\right)^2 + \left(\frac{T_c}{6}\right)^2$$

or:

$$\left(\frac{0.026}{6}\right)^2 = \left(\frac{0.006}{6}\right)^2 + \left(\frac{0.008}{6}\right)^2 + \left(\frac{0.008}{6}\right)^2 + \left(\frac{2X}{6}\right)^2$$

$$X = 0.011$$

The resulting tolerance is much larger (and much less costly to achieve) than the tolerance that was found by demanding 100% interchangeability. On the negative side, if the component dimensions are normally distributed, then 100% - 99.73% or 27 out of 10,000 assemblies should be expected to be out of specification.

13.5 Qualifying Tolerances with Gages

Quality control of manufactured components often calls for determining if a dimension is within the limits specified by the tolerance. This type of inspection is performed with *gages*, special tools designed to determine if a fixed dimension lies within the proper limits. A gage is dedicated to a particular dimension and specific tolerances. If a dimension or the associated tolerances are changed a new gage must be made.

Two types of gages are commonly used for qualifying parts in mass production, plug gages and snap gages. A snap gage, as shown in Figure 13.6(a), is used for external dimensions such as the diameter of a shaft . A plug gage, as shown in Figure 13.6(b), is used to qualify an internal dimension, typically an internal diameter. Both of these gages have a go and a no-go feature on them. In the case of a plug gage, one side must be able to "go into" a hole being gaged while the other side must "not go into" the same hole for the diameter to be within the specified limits.

The key principle that is applied in the design of any gages states: *it is better to reject a good part than declare a bad part to be within specifications.* All gage design decisions are made with this principle in mind. Gages must have tolerances like any manufactured components. All gages are made with:

• **Gage Tolerance.** This tolerance states the permissible variation in the manufacture of the gage. It is typically 5% to 20% (depending on the industry) of the tolerance on the dimension being gaged.

• **Wear Tolerance.** This tolerance compensates for the wear of the gage surface as a result of repeated use. Wear tolerance is only applied to the "go" side of the gage since the "no-go" side should seldom see contact with a part surface. It is typically 5% to 20% of the dimensional tolerance.

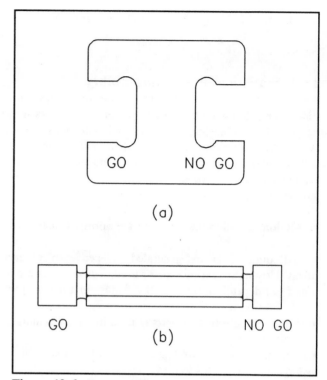

Figure 13.6 Snap and Plug Gages.

The following examples will illustrate the design of gages to qualify diameter dimensions. *Plug gages* are used to qualify internal diameters. *Snap gages* are used to qualify external diameters.

Example 13.5.1. Design a gage to qualify the internal diameter shown in Figure 13.7. Use 20% wear and gage tolerances.

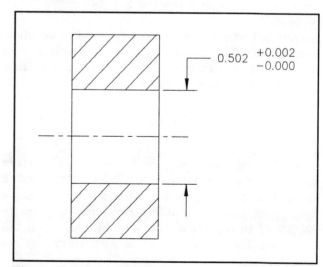

Figure 13.7 Internal Diameter to be Qualified.

Solution. The gage and wear tolerance are determined first. The total gage tolerance will be 20% of the total tolerance of the dimension. In this case:

$$0.2 \times 0.002 = 0.0004$$

This is distributed evenly on both the go and no-go sides of the gage. Each side of the gage has a 0.0002-inch tolerance applied unilaterally. The total wear tolerance will be 20% of the total tolerance on the dimension of 0.0004 inch. This is distributed evenly on both sides of the gage, but it will only be applied to the go side, which is typically the only side of the gage subject to wear. The go side has a 0.0002-inch wear tolerance applied to the diameter of the "plug" or diameter used for gaging.

Next, the go and no go side plug dimensions are determined.

Go Side—The go side must be inserted into the smallest hole that meets specifications (0.500 inch). The gage dimension is based on this diameter increased by one half the total wear tolerance so that the gage diameter approaches the smallest specified diameter as the gage wears. Some parts that are within specification will be rejected, but this is consistent with the principle of gage design. The gage tolerance is similarly applied. The 0.0002-inch tolerance is applied unilaterally on the positive side. This permits a gage to be fabricated that will reject a small number of good parts, but this condition is consistent with the design principle.

No-go Side—The no-go side *must not* be able to be inserted in any hole within specifications. The nominal diameter of this side is equal to the upper limit on the gaged dimension or 0.502 inch. The 0.0002-gage tolerance is applied unilaterally on the negative side. Consequently, it is possible that the no-go side can be inserted into a small number of holes that are within specification, but this is consistent with the design principle.

The resulting gage design is shown in Figure 13.8.

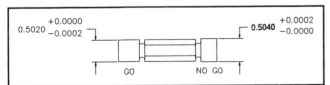

Figure 13.8 Plug Gage Design.

Example 13.5.2. Design a gage to qualify an external diameter on a shaft as shown in Figure 13.9. Use 20% wear and gage tolerances.

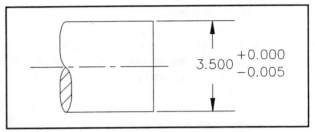

Figure 13.9 External Diameter to be Qualified.

Solution. Gage Tolerance—(0.2)(0.005 inch) = 0.0010 inch. Each side of the gage has a 0.0005-inch gage tolerance applied unilaterally to the distance between the gage surfaces.

Wear Tolerance—(0.2)(0,.005 inch) = 0.0010 inch. Since the go side is subject to the most wear, a 0.0005-inch wear tolerance is applied to the gap distance on the go side.

Go Side—The go side must be able to fit over the largest shaft within specifications (3.500 inches). This diameter is decreased by one half of the total wear tolerance so that the gage approaches the largest specified diameter as the gage wears. The 0.0005-inch gage tolerance is applied unilaterally on the negative side.

No-go Side—The no-go side should not be able to fit over any shaft within specifications. The nominal diameter of this side is equal to the lower limit on the gaged dimension or 3.495 inches. The 0.0005-inch gage tolerance is applied unilaterally on the positive side.

The resulting gage design is shown in Figure 13.10.

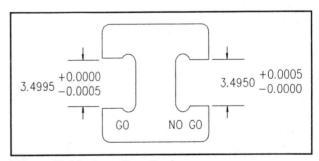

Figure 13.10 Snap Gage Design.

Practice Problems

13.1 A dowel pin is toleranced 1.0000 ± 0.0001. Give the required tolerances for a matching hole with 0.0001 allowance.

13.2 A dimension may vary between 0.505 and 0.508. Express this information in the form of a limit dimension, a unilateral tolerance and a bilateral tolerance.

13.3 Which method of tolerancing will generate larger tolerances, 100% interchangeability or statistical interchangeability?

13.4 Provide proper dimensions and tolerances for a piston and a cylinder according to the following specifications: The nominal dimension of the piston is 1.000 inch. The allowance in the assembly is 0.001 inch. The piston and cylinder each have total, unilateral tolerances of 0.001 inch.

13.5 Find tolerance X in Figure P13.1 for 100% interchangeability.

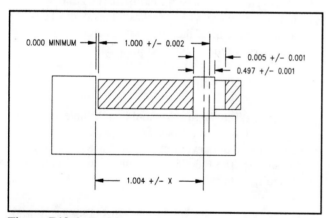

Figure P13.1 Problem 13.5.

13.6 Find tolerance X in Figure P13.1 for statistical interchangeability.

13.7 Find tolerance Y in Figure P13.2 for 100% interchangeability.

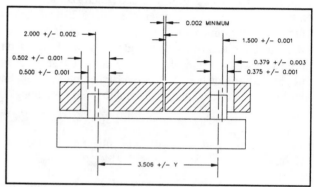

Figure P13.2 Problem 13.7.

13.8 Find tolerance Y in Figure P13.2 for statistical interchangeability.

13.9 Design a go/no go gage capable of qualifying a shaft with the following dimension:

$$2.150 \, ^{+0.000}_{-0.005} \quad inch$$

Use 10% wear and gage tolerances.

13.10 Design a go/no go gage capable of qualifying the following internal diameter.

$$3.000 \, ^{+0.010}_{-0.010} \quad inch$$

Use 10% wear and gage tolerances.

Part 5

Engineering Management

QUALITY ASSURANCE

Quality assurance refers to the effort undertaken by a manufacturer to ensure its products conform to specifications. The specifications for a product are intended to guarantee the proper assembly of components, freedom from defects and intended performance. Quality assurance is the responsibility of everyone involved in the design and manufacturing processes associated with a product. However, this section will be limited to the techniques used to *observe* a process and detect the potential for the production of defects. Many of these techniques can be categorized as *Statistical Quality Control*.

14.1 Inspection

Inspection is the process of checking the conformance of a final product to its specifications. In most cases, 100% inspection of a process is too costly. Therefore, there are established methods of *sampling* a product or process to characterize its correspondence to its specifications. Inspection must be a continuous activity because raw material, machines and operators are all subject to *variability*.

Two types of inspection are typically employed in a quality assurance activity. Inspection of *variables* involves the quantitative measurement of characteristics such as dimensions, surface finish and other physical or mechanical properties. Such measurements are made with instruments that produce a variable result. A micrometer or a thermometer are typical tools used in inspection of variables. The resulting measurements are compared against specifications and conformance assessments are made. The other type of inspection is *attribute*. This approach involves the direct observation of a quality characteristic. A GO/NO GO gage offers the direct comparison of a dimension to a specification. The presence or absence of a flaw, such as a visible scratch in a painted panel, is another direct observation of a quality characteristic.

14.2 Statistical Methods for Quality Control

Statistical methods refers to extracting the significant information from large amounts of numerical information. This approach is important in quality control since large quantities of material or product may be involved. Statistical methods are also employed when dealing with variability in data, such as in manufacturing processes. No two products are ever manufactured exactly alike. There are always variations in the dimensions or properties of raw materials, and variation in the operation of machines and operator performance. Statistics are also important tools for quality assurance because they offer a way of characterizing a *population* (all of the individual parts) by means of a *sample* (a small group of parts studied).

One of the most important results of statistical quality control (SQC) is to detect variation in the process. There are two types of variation occurring in a process, *natural* and *assignable*. *Natural variability* in a manufacturing process is the inherent, uncontrolled changes that occur in the composition of material, the performance of the operator and the operation of machines. These variations occur *randomly* with no particular pattern or trend. In contrast, *assignable variability* can be traced to a specific, controllable cause. SQC methods are intended to distinguish between natural and assignable variability. Ideally, if the assignable causes of variability can be identified, the process can be better controlled and defects can be prevented. The systematic method of detecting assignable variability in a process is known as *statistical process control* (SPC).

To use SPC tools, certain concepts from statistics must be employed. Assume, for example, that a series of several dozen shafts are being turned on a lathe. A micrometer will be used to measure the diameter. After a short period, it becomes evident that the diameters vary. If they were listed in order from smallest to largest, it could be seen that very few diameters are close to the two extremes. Most would lie between the two extremes with fewer diameters found near the extreme small or large size.

If the diameters were collected into groups defined by a minimum and maximum size, the size of these groups would indicate how many shafts had diameters close to a particular dimension. If these groups were plotted on a bar graph with the diameter on the horizontal axis and the size of the group on the vertical axis, a *distribution* of the shaft diameters would be revealed as in Figure 14.1. The distribution describes how often a particular diameter occurs.

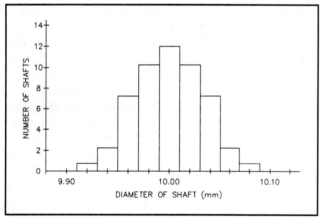

Figure 14.1 Distribution of Shaft Diameters.

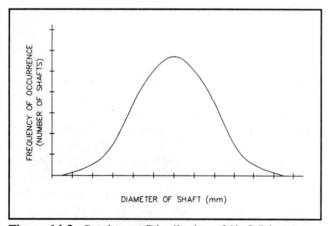

Figure 14.2 Continuous Distribution of Shaft Diameters.

If a very large number of measurements are made and the width of the bars in the graph are made very thin, the frequency of occurrence of a diameter would follow a characteristic "bell-shaped" distribution as shown in Figure 14.2. Data from many manufacturing processes (and a large number of other naturally occurring processes) has this characteristic distribution. This distribution corresponds well to a theoretical distribution known as the *normal distribution* (also called a *gaussian distribution*). The shape of this distribution is repeatable, there are only two

parameters needed to completely characterize it. These two parameters are the *mean* (or average value) and the *dispersion*. Figure 14.3 shows several different normal distributions with different means and dispersions.

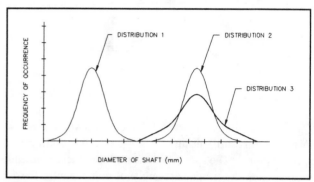

Figure 14.3 Normal Distributions with Different Means and Dispersions.

The distributions illustrated in Figure 14.3 show the significance of the mean and dispersion. Distribution 1 has the same dispersion as Distribution 2, however, their measurements are clustered around different mean values. Distribution 2 has the same mean as Distribution 3, but the dispersion of Distribution 3 is much greater, indicating that the measurements tended to vary more from the mean value.

The mean shows the value that the measurements tend to cluster around. The *arithmetic mean* is often called the average value $\bar{x}$ (called "x bar") and can be estimated using the following familiar formula:

$$\bar{x} = \frac{x_1 + x_2 + x_3 \ldots + x_n}{N}$$

where the numerator is the sum of all measurements and the denominator N is the number of measurements. The dispersion is the width of the normal curve and indicates how much variability is present in a set of measurements. Many different measures of dispersion can be used. One of the simplest is the range R which is the difference between the largest and smallest measured value:

$$R = x_{max} - x_{min}$$

Another measure of dispersion is the *sample standard deviation* σ which can be estimated by:

$$\sigma = \sqrt{\frac{(x_1 - \bar{x})^2 + (x_2 - \bar{x})^2 + (x_3 - \bar{x})^2 + \ldots + (x_n - \bar{x})^2}{n - 1}}$$

124

The units standard deviation are the same as the units of the original measurements. In the lathe turning example, standard deviation would be expressed in millimeters.

Example 14.2.1. Calculate the mean and sample standard deviation of the group of shaft diameters shown in Table 14.1.

Table 14.1 Observed Shaft Diameters.

Observation	Diameter (mm)	Observation	Diameter (mm)
1	1.05	5	1.03
2	0.96	6	1.02
3	0.99	7	0.99
4	1.04	8	1.01

Solution. The mean is found by summing the measurements and dividing by the number of measurements:

$$\bar{x} = \frac{1.05+0.96+0.99+1.04+1.03+1.02+0.99+1.10}{8} = 1.01$$

The standard deviation is found by taking the square root of the sum of the squares of the difference between each measurement and mean and divided by the number of measurements minus one:

$$\sigma = \sqrt{\frac{(1.05-1.01)^2+(0.96-1.01)^2+...+(1.01-1.01)^2}{8-1}} = 0.03$$

One of the most important uses for the mean and standard deviation in a normal distribution is predicting the percentage of measurements that will fall into a certain range. Figure 14.4 shows the percentage of measurements that fall into ranges defined by distance measured in number of standard deviations for a normal distribution. For example, 68.26% of all observations will fall within plus or minus one standard deviation from the mean.

Example 14.2.2. The mean diameter of a part is 25.00 mm and the standard deviation is 0.02 mm. The process is normally distributed. Estimate how many parts will have a diameter greater than 25.04 mm.

Solution. The diameter 25.04 mm is two standard deviations greater than the mean:

$$25.04 = \bar{x} + 2\sigma = 25.00 + 2(0.02)$$

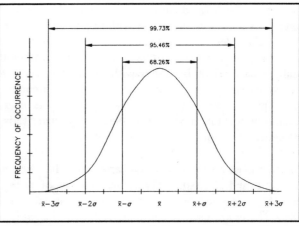

Figure 14.4 Percentage of Occurrence in a Normal Distribution.

The percentage of parts in question is shown as the area under the bell-shaped curve in Figure 14.5. The percentage of parts less than $\bar{x} - 2\sigma$ and greater than $\bar{x} + 2\sigma$ can be found by:

$$100\% - 95.46\% = 4.54\%$$

The number of diameters greater than $\bar{x} + 2\sigma$ is half this amount or 2.27% due to the symmetry of the normal curve about the mean.

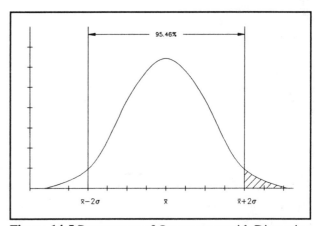

Figure 14.5 Percentages of Occurrences with Dimension Greater Than the Mean Plus Two Standard Deviations.

14.3 Statistical Process Control

One of the principal tools used in SPC is the *control chart*. Control charts are a plot of a quality characteristic (such as the average diameter of a sample of parts or the number of defects in a sample) with respect to time. The quality characteristic is compared to *control limits*. The basic purpose for the chart is to determine whether the quality characteristic is varying within acceptable limits for natural variability or whether the process is "in control." The typical control charts display limits for the natural variability for the process and the quality characteristic calculated at various times.

The two most common control charts in use for *variables* are the $\bar{x}$ and R charts. (There are other types of charts used for attributes, but the basic concept is the same.) These charts are used to show the statistics of *samples* of measurements made on the process. The $\bar{x}$ chart shows the quality characteristic of the mean of the sample. The R chart shows the quality characteristic of the range of the sample. The $\bar{x}$ and R charts are typically plotted together and used together to interpret the performance of a process. A set of typical $\bar{x}$ and R charts are shown in Figure 14.6.

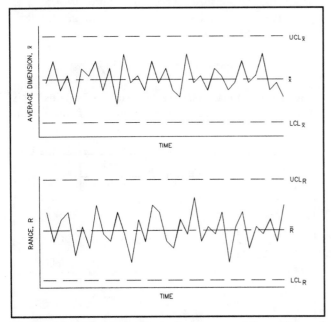

Figure 14.6 Control Charts for Average Dimension and Range.

The concept of a *sampling* is used to efficiently represent the overall performance of the process with a limited number of observations. A sample size of 2 to 20 is typically used for $\bar{x}$ and R charts. This means that a fixed quantity of randomly selected parts is used to estimate the quality characteristics for corresponding to a particular period of time. For example, five shafts might be selected every hour for generating the control chart data. The

diameters of the five shafts are averaged to obtain the $\bar{x}$ characteristic for that hour and the difference between the largest and smallest diameter are used to obtain the R characteristic. In general terms, $\bar{x}$ characterizes the typical measurement and R characterizes the variability of the measurement. If the process is in statistical control, the typical measurement and the variability of the measurement will change somewhat over time, but these changes should be random and limited in their magnitude. If the changes in $\bar{x}$ and R occur in a pattern or have excessive magnitude, there is assumed to be an assignable cause. Such pronounced changes are not typically part of the inherent variability in the process. Figure 14.7 shows an $\bar{x}$ chart that exhibits control for the first 10 samples, but then shows a marked shift in magnitude. This type of change is almost always assigned to a cause such as changing a tool, adjusting a machine, etc. Such assignable causes can be controlled, thus the performance of process can be improved by some type of intervention.

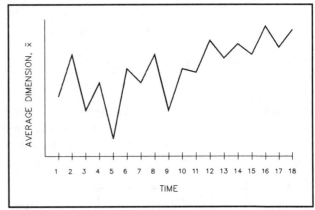

Figure 14.7 Deviation of the Average Dimension in a Control Chart.

A control chart needs some limits that define the maximum acceptable deviations of the quality characteristic for a process in statistical control. These *control limits* are established when the process is under close observation and known to be operating properly. The control limits are used as a standard to compare future performance. Both $\bar{x}$ and R have average values that normal variances in the process fluctuate about. Figure 14.8(a) shows a horizontal line labelled $\bar{\bar{x}}$. This is the *average of averages* or *grand average*. It represents the average of the typical values for a measurement and represents the population mean. Figure 14.8(b) shows a horizontal line labelled $\bar{R}$. This represents the typical variance in the process. These two parameters provide the basis for determining the control limits on a control chart.

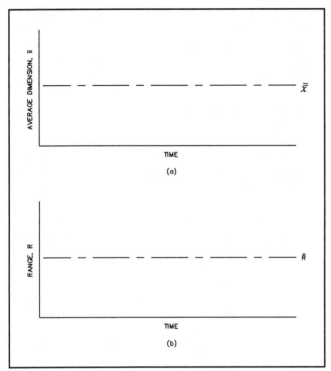

Figure 14.8 Mean Lines in Control Charts.

Example 14.3.1. A shaft diameter is measured as part of a SPC plan. Four observations of the diameter are recorded in each subgroup (labelled x_1 – x_4, as shown in the table below). Calculate $\bar{\bar{x}}$ and $\bar{R}$.

Sample	x_1	x_2	x_3	x_4	$\bar{x}$	R
1	5.005	4.994	5.002	5.004	5.001	0.011
2	4.998	4.999	5.005	5.001	5.001	0.007
3	5.002	5.006	5.000	4.999	5.002	0.007
4	5.003	5.005	5.000	4.997	5.001	0.008
5	5.001	4.996	4.995	5.005	4.999	0.010
				Sum	25.004	0.043

Solution. The quantity $\bar{x}$ is found by averaging the four measurements in each sample. The quantity R is found by computing the difference between the largest and smallest measurement in each sample. The sum of each $\bar{x}$ is used to compute $\bar{\bar{x}}$:

$$\bar{\bar{x}} = \frac{25.004}{5} = 5.0008 \approx 5.001$$

The sum of each R is used to compute $\bar{R}$:

$$\bar{R} = \frac{0.043}{5} = 0.0086 \approx 0.009$$

The quality characteristics $\bar{x}$ and R tend to be normally distributed (for reasons related to an idea from statistics known as the *Central Limit Theorem*). Consequently, most of the variation in $\bar{x}$ and R can be characterized by three standard deviations of the quality characteristic from the average quality characteristic. The upper control limit on the $\bar{x}$ chart, $UCL_{\bar{x}}$, is given by:

$$UCL_{\bar{x}} = \bar{\bar{x}} + 3\,\sigma_{\bar{x}}$$

Note that the standard deviation used is for $\bar{x}$, not simply for x. The control limit characterizes the variability of the typical performance of the process, not the variability in the population. Similarly, the lower control limit is given by:

$$LCL_{\bar{x}} = \bar{\bar{x}} - 3\,\sigma_{\bar{x}}$$

The computation of the standard deviation of $\bar{x}$ is sometimes a time consuming and error prone operation. In production operations, a simplified means of estimating the control limits on an $\bar{x}$ chart are typically used:

$$UCL_{\bar{x}} = \bar{\bar{x}} + 3\,\sigma_{\bar{x}} = \bar{\bar{x}} + A_2\bar{R}$$

$$LCL_{\bar{x}} = \bar{\bar{x}} - 3\,\sigma_{\bar{x}} = \bar{\bar{x}} - A_2\bar{R}$$

where A_2 is a variable based only on sample size. This means of estimating control limits is accurate enough for virtually all applications. This approach has been proven to be less error prone than one that calls for calculating $\sigma_{\bar{x}}$ directly. The constant A_2 is given for various sample sizes in Table 14.2.

Table 14.2 Constants for Control Charts

Sample Size	A_2	D_4	D_3	d_2
2	1.880	3.267	0	1.128
3	1.023	2.575	0	1.693
4	0.729	2.282	0	2.059
5	0.577	2.115	0	2.326
6	0.483	2.004	0	2.534
7	0.419	1.924	0.078	2.704
8	0.373	1.864	0.136	2.847
9	0.337	1.816	0.184	2.970
10	0.308	1.777	0.223	3.078

The second control chart, seen in Figure 14.6, shows the range in each sample of observations. As stated earlier,

127

this chart characterizes the variability of the measurements. The line in the center of the chart, $\bar{R}$, is actually a measure of the variability in the process. The upper and lower control limits should bound the natural deviations in the range in each sample. Since the range can be treated as a normally distributed phenomena (according to the Central Limit Theorem), the control limits can be formed as:

$$UCL_R = \bar{R} + 3\sigma_R$$

$$LCL_R = \bar{R} - 3\sigma_R$$

For simplicity, the control limits can be estimated by:

$$UCL_R = D_4 \bar{R}$$

$$LCL_R = D_3 \bar{R}$$

where the constants D_3 and D_4 are functions of the sample size and are given in Table 14.2.

Table 14.2 also includes a constant d_2 which is useful for estimating the standard deviation of the process σ:

$$\hat{\sigma} = \frac{\bar{R}}{d_2}$$

the caret or hat above the standard deviation indicates that it is an estimate.

Example 14.3.2. Find the control limits for the $\bar{x}$ and R charts using the data from Example 14.3.1.
Solution. Since the sample size is four, the following constants are found in Table 14.2:

$$A_2 = 0.729, \ D_3 - 0 \text{ and } D_4 = 2.282$$

The results of Example 14.3.1 can be applied to determine the control limits for the $\bar{x}$ chart:

$$UCL_{\bar{x}} = \bar{x} + A_2 \bar{R} = 5.001 + (0.729)(0.009) = 5.008$$

$$LCL_{\bar{x}} = \bar{x} - A_2 \bar{R} = 5.001 - (0.729)(0.009) = 4.994$$

Similarly, the control limits on the R chart can be found as:

$$UCL_R = D_4 \bar{R} = (2.282)(0.009) = 0.021$$

$$LCL_R = D_3 \bar{R} = (0)(0.009) = 0$$

The most important use for control charts is tracking down assignable causes for changes in process performance as part of a process improvement activity. Figure 14.11(a), page 130, shows an $\bar{x}$ control chart that exhibits good statistical control. There is no obvious trend in the data. The sample averages are random with time and all points are contained within control limits. An obvious case where an assignable cause might be identified is illustrated in Figure 14.11(b). The sample averages clearly exceed the upper control limit. An investigation into the circumstances that surrounded the process at the time of the control limit excursion might reveal that a cutting tool was changed. A shift in level, even one that does not exceed the control limit, can be an indicator of nonrandom behavior, as shown in Figure 14.11(c). This type of change might be attributable to changing the supplier for the incoming material used in the process. Finally, a repeated pattern is another form of nonrandom behavior. Figure 14.11(d) shows a cyclic pattern in the sample average. This type of performance may be related to shop temperature (cold in the morning and warm in the afternoon). Possibly the low and high points are caused by a machine or operator that behaves differently when the temperature is high or low.

14.4 Process Capability Analysis

Process capability analysis is a method of determining or assuring that a process can meet specifications. A process is said to be *capable* if it will be able to consistently produce parts within specification. The capability of a process can be viewed as the relationship between the specified limits for a dimension and the limits of the natural variability of the dimension. Figure 14.9 shows a frequency distribution for a dimension. The upper

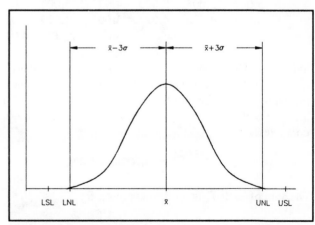

Figure 14.9 Relationship of Specified Limits and Natural Limits for a Capable Process.

specified limit and the lower specified limit (possibly taken from an engineering drawing) are designated as the USL and LSL, respectively. There are two additional limits shown, the lower natural limit (LNL) and the upper natural limit (UNL). These limits reflect 3σ variation on either side of the mean dimension. If the process is capable, the natural limits, which describe the typical range of variation for a measurement, will fall within the specified limits.

The extent to which the natural limits fall within the specified limits is a variable measure of process capability. The *Process Capability Ratio* (often designated as C_p) is one measure that is frequently used for processes that have natural limits centered within the specified limits. The Process Capability Ratio can be calculated by dividing the difference in the specified limits by the total (or almost total) amount of variation expected in process:

$$C_p = \frac{USL - LSL}{6\sigma}$$

where 6σ provides a measure of the variation in the process. Note that σ is used and not $\sigma_{\bar{x}}$.

Clearly, C_p should be greater than or equal to one or else a significant percentage of defective parts will be produced. If C_p is less than one, as shown in Figure 14.10, the process is generally considered incapable of producing good parts. A larger value of C_p is preferable. Ratios above 1.33 are considered acceptable. This ratio can be used to quantify and compare process capability. Most major manufacturers insist that their suppliers provide a report of the C_p for a process. This variable is often used as an indicator of a good supplier.

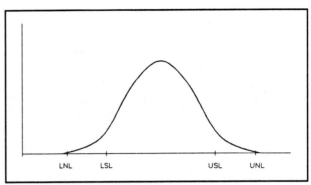

Figure 14.10 Incapable Process.

Example 14.4.1. If the specifications on the dimension being monitored in Example 14.3.2 are:

$$5.000^{+0.020}_{-0.020}$$

find the Process Capability Ratio.

Solution. The process standard deviation needs to be found. The average range of the process is used in conjunction with the constant d_2 from Table 14.2 for a sample size of four to find:

$$\sigma \approx \hat{\sigma} = \frac{\bar{R}}{d_2} = \frac{0.009}{2.059} = 0.004$$

Then C_p can be found as follows:

$$C_p = \frac{USL-LSL}{6\sigma} = \frac{5.020 - 4.980}{6\,(0.004)} = 1.667$$

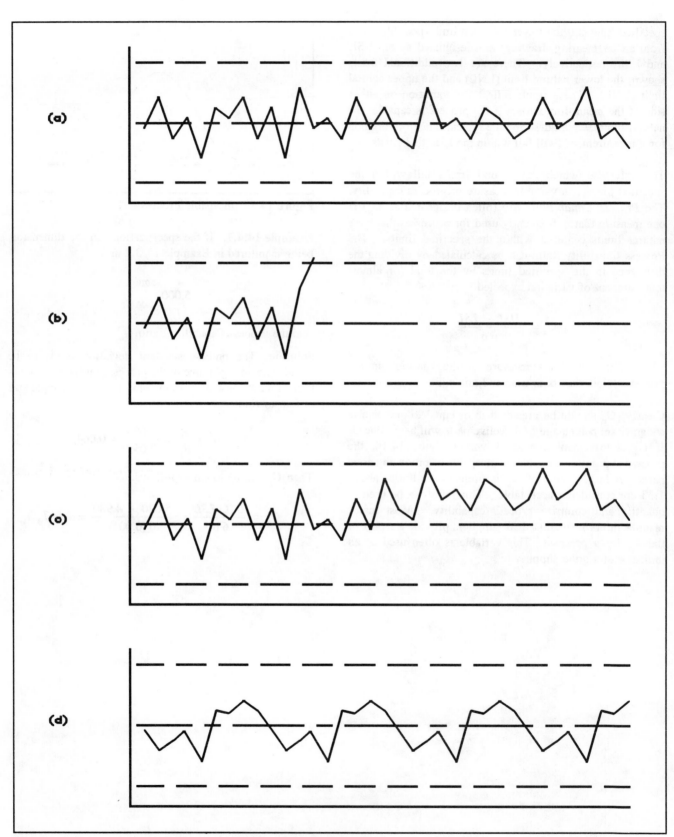

Figure 14.11 Various Conditions Being Monitored on a Control Chart.

Practice Problems

14.1 Give two examples of *variables* and two examples of *attributes* used in inspection.

14.2 Give two examples of *assignable variability* and *natural variability* in the machining of parts on a lathe.

14.3 Give two measures of dispersion.

14.4 Find the mean and standard deviation of the following set of observations: {0.251, 0.248, 0.252, 0.253, 0.250, 0.249, 0.249, 0.251}.

14.5 The mean diameter of a bored hole is 1.875 inches. The standard deviation of the diameter is 0.0005 inch. A total of 1000 parts is to be bored. Estimate the number of parts with diameters smaller than 1.8745 inches.

14.6 Calculate the limits for x-bar and range control charts using the data in Table P14.1.

14.7 Calculate the limits for x-bar and range control charts using the data in Table P14.2.

14.8 Estimate the standard deviation of the process being monitored in Problem 14.6.

14.9 The tolerance on the dimension being monitored in Problem 14.7 is specified as:

$$21.00 \, {}^{+2.00}_{-2.00}$$

Find the process capability ratio. Is the process capable of producing an acceptable percentage of good parts.

Table P14.1 Problems 14.6 and 14.8.

Subgroup	Observations made on parts in Subgroup				
	A	B	C	D	E
1	77	80	78	72	78
2	76	79	73	74	73
3	76	77	72	76	74
4	74	78	75	77	77
5	80	73	75	76	74
6	78	81	79	76	76
7	75	77	75	76	77
8	79	75	78	77	76
9	76	75	74	75	75
10	71	73	71	70	73
11	72	73	75	74	75
12	75	73	76	73	73

Table P14.2 Problems 14.7 and 14.9.

Subgroup	Measurements in Subgroup			
1	22.0	22.5	22.5	24.0
2	20.5	22.5	22.5	23.0
3	20.0	20.5	23.0	22.0
4	21.0	22.0	22.0	23.0
5	22.5	19.5	22.5	22.0
6	23.0	23.5	21.0	22.0
7	19.0	22.0	22.0	20.5
8	21.5	20.5	19.0	19.5
9	21.5	22.5	20.0	22.0
10	21.5	23.0	22.0	23.0
11	20.0	19.5	21.0	20.0
12	19.0	21.0	21.0	21.0
13	19.5	20.5	21.0	20.5

ENGINEERING ECONOMICS

Engineering economics is the name given to techniques for evaluating financial decisions in the engineering enterprise. The objective is to provide a means of making economically sound decisions in the execution of engineering projects.

15.1 Time Value of Money

Over a short period of time, sums of money can be treated in the same way as any other algebraic quantity. In the short term, simple addition and subtraction is all that is necessary to evaluate economic alternatives. However, money cannot be treated this way over longer periods of time. Money available today is always worth more than the same amount of money available at some time in the future. Consequently, a thousand dollars today is more valuable than a guarantee of a payment of a thousand dollars a year from now. We could deposit the $1000 in a bank and at the end of a year period, we could collect both the $1000 and the interest that was earned. If the interest rate is 5%, $1000 today is really equivalent to $1050 in a year from now. Clearly, economic decision-making requires careful consideration of the value of money over time.

15.2 Cash Flow Patterns

Engineering economics requires the comparison of patterns of cash flow. A car loan that requires a down payment and a series of equal monthly payments is an example of a pattern of cash flow. This review will consider three patterns of cash flow:

- **P-pattern** A single amount P occurs at the beginning of n periods. P represents the principal or present amount. This quantity might refer to a single deposit in a mutual fund for later use.

- **F-pattern** A single amount F occurs at the end of n periods. F represents the future amount. This quantity might represent the withdrawal from a long-term savings plan for retirement.

- **A-pattern** Equal amounts of A occur at the ends of n periods. A represents an annual amount (although the period may be a month or other period). This quantity might represent the payment made on a mortgage.

The solution to most engineering economics problems involve finding a pattern of cash flow equivalent to another pattern of cash flow. For example, how much money must be deposited in a bank at a given interest rate to yield a desired amount in the future. This problem can be thought of as finding the amount in an F-pattern that is equivalent to a P-pattern. These two amounts are proportional with a proportionality factor that is dependent on the interest rate per period i and the number of interest periods n. The number of periods depends on the frequency of compounding (often this occurs monthly).

There are symbols for the proportionality factors that have an appearance that suggests algebraic cancellation. This notation is designed to prevent selection of the wrong factor in a given problem. In the case of determining the amount that must be deposited now to yield a desired sum in the future, the proportionality factor is written $(P/F,i,n)$. The factor is applied in the following equation:

$$F = (F/P,i,n)\, P$$

This indicates that a present sum of P is required at an interest rate of i for n periods given a desired future sum of F. Table 15.1 illustrates some various constants of

Table 15.1 Proportionality Factors for Compound Interest Formulas.

Symbol	To Find	Given	Formula	Name and Example
(F/P,i,n)	F	P	$(1 + i)^n$	Single Payment Compound Amount Factor (Find the amount that results from leaving a given amount in a bank account.)
(P/F,i,n)	P	F	$\dfrac{1}{(1 + i)^n}$	Single Payment Present Worth Factor (Find the amount that must be left in a bank account in order to yield a desired amount.)
(A/P,i,n)	A	P	$\dfrac{i(1 + i)^n}{(1 + i)^n - 1}$	Uniform Series Capital Recovery Factor (Find the payment on a car loan.)
(P/A,i,n)	P	A	$\dfrac{(1 + i)^n - 1}{i(1 + i)^n}$	Uniform Series Present Worth Factor (Find the amount that can be borrowed given a fixed monthly payment.)
(A/F,i,n)	A	F	$\dfrac{i}{(1 + i)^n - 1}$	Uniform Series Sinking Fund Factor (Find the amount that must be deposited into an IRA each month to yield a desired amount at retirement.)
(F/A,i,n)	F	A	$\dfrac{(1 + i)^n - 1}{i}$	Uniform Series Compound Amount Factor (Find the amount that results from a series of fixed deposits into a bank account.)

proportionality and the formulas used to find them.

The numerical values for the factors in Table 15.1 can also be found in tabular form. Many texts on engineering economics provide these factors in this form as an alternative way of solving these types of problems.

Example 15.2.1. How long will it take for $10,000 invested in a bank savings account to double in value if the bank pays 6% interest compounded quarterly?
Solution. The problem is one where the future sum F ($20,000) must be generated based on the present sum P ($10,000). There are four periods per year with a periodic interest rate of 1.5%. The unknown in the problem is the number of periods. The problem is represented by:

$$F = (F/P,0.015,n)\, P$$

For the future sum to be twice the principal investment, the equation is written:

$$2 = (F/P,0.015,n)\, 1$$

or

$$2 = (1 + 0.015)^n$$

$$\ln 2 = n \ln (1.015)$$

$$n = 46.6 \approx 47$$

Thus it will require 47 quarter-year periods or just under 11 3/4 years for the investment to double.

Example 15.2.2. A truck is to be purchased for $78,000. The truck dealer offers terms of a $5000 down payment with 12 monthly payments at an annual rate of 12%. What is the monthly payment?
Solution. The problem required finding the "annual sum" A (although in this case the payments are monthly) based on the present sum P of $73,000 (the principal of the loan). The number of periods is 12 with a periodic interest rate of 1% (12/12%). The problem can be represented by the equation:

$$A = (78000 - 5000)\,(A/P,0.01,12)$$

$$= 73000\, \frac{0.01\,(1.01)^{12}}{(1.01)^{12} - 1}$$

$$= 73000(0.0888) = \$6482$$

Example 15.2.3. The goal of a savings plan is to accumulate $10,000 at the end of 10 years. How much money must be invested now if the savings account offers 4% interest, compounded quarterly?

Solution. The problem requires finding the present sum based on a known future sum. The period interest rate is 1% over 40 periods. This can be expressed as:

$$P = F(P/F,0.01,40) = 10000\,(P/F,0.01,40)$$

$$= 10000\,(0.6717) = \$6717$$

Example 15.2.4. What amount must be deposited in a bank at 5% interest, compounded annually, to provide $1000 per year for the next 50 years?

Solution. The problem requires finding a present sum based on a known annual sum.

$$P = 1000\,(P/A,0.05,50) = 1000\,(18.256) = \$18256$$

15.3 Comparisons Based on Annual Cost

There are many different techniques used in engineering economic analysis to evaluate alternatives. One of the most common is comparison of equivalent uniform annual cost (EUAC). This technique allows the comparison of a nonuniform series of cash flows to identify the minimum cost alternative. This approach is used in deciding between two alternative investments in equipment, property or other resource, known as an *asset*.

There are several assumptions made in applying this approach:

• There is a uniform time value or interest rate on all money involved in the problem whether it is borrowed or not. Money that is not invested represents an opportunity cost of lost interest.

• The annual cost of an asset is reduced by the money made from the sale or salvage of an asset at the end of its useful life.

• If two alternatives have different useful service life, it is assumed the asset with the shorter life, will be replaced with an identical item.

The solution to these types of problems requires identifying all the components of the annual cost including the opportunity cost of not investing the present value of the asset, the operating cost and the cost reduction associated with the salvage of the asset.

Example 15.3.1. A company is considering purchasing a machine for $10,000. After 12 years of use, there is a projected salvage value of $3,000. The machine will require $150 per year in maintenance. Determine the equivalent uniform annual cost (EUAC) if the interest rate is taken as 10%.

Solution. The EUAC has three components: the annual opportunity cost of the purchase price, the maintenance cost and the equivalent annual benefit of the salvage value of the machine.

$$EUAC = 10000\,(A/P,0.1,12) + 150 - 3000\,(A/F,0.1,12)$$

$$= 10000\,(0.1468) + 150 - 3000\,(0.0468) = \$1479$$

Example 15.3.2. A used delivery van can be purchased for $18,000. Its salvage value is estimated to be $4,000 after a seven-year service life. The daily operating cost of the truck is $125 including fuel, maintenance and the driver's salary. A similar service can be obtained from a delivery service for $160 per day. How many days per year must the truck be used to justify its purchase? Assume an 8% interest rate.

Solution. Let the number of days the truck is used per year be x. Set the annual cost of hiring the delivery service be equal to the annual cost of owning the truck.

$$160x = 18000\,(A/P,0.08,7) + 125x - 4000\,(A/F,0.08,7)$$

$$35x = 18000\,(0.1921) - 4000\,(0.1121)$$

$$x = 86\,days$$

Example 15.3.3. A specialized machine tool costs $500,000. It has an estimated life of 20 years. What amount should the company be willing to spend on *extra* maintenance if it would extend the service life to 30 years? Assume a 12% interest rate.

Solution. The company should only invest in extra maintenance (EM) until the EUAC of the 30-year service life program is equal to the EUAC of the 20-year service life program. Therefore, the problem can be expressed as:

EUAC for 30 year service = EUAC for 20 year service

$$500000\,(A/P,0.12,30) + EM = 500000\,(A/P,0.12,20)$$

$$500000\,(0.1241) + EM = 500000\,(0.1339)$$

$$EM = \$4900$$

Practice Problems

15.1   If a savings plan has a goal to accumulate $10,000 after 12 years of annual deposits. An annual interest rate of 8% is available. What should the annual deposit be?

15.2   A car is to be purchased for $15,000. A $4000 down payment is made. The remainder is borrowed at an annual interest rate of 12%. Find the monthly payment for a three-year loan.

15.3   If the goal of a savings plan is to accumulate $50,000 after 20 years, how much should be deposited now if an annual interest rate of 6% is available?

15.4   How much money should be deposited now to generate $20,000 per year for the next 10 years. An 8% interest rate is available.

15.5   A new machine with a life of six years is expected to save $2000 in production coists each year. What is the highest price that can be justified for the machine? Use a 12% annual interest rate.

15.6   A machine costs $46,000 and has an expected life of five years. A service contract costing $4000 per year is available which promises a 7-year service life. If a 9% interest rate is available, is the service contract a sound economic decision?

15.7   A machine is to be purchased for $40,000. The expected life of the machine is five years. Its anticipated salvage value is $10,000. The annual maintenance cost is $1200. Find the equivalent uniform annual cost based on a 12% annual interest rate.

15.8   An additional computer can be purchased to process payroll checks for $15,000. It has an estimated salvage value of $2000 after a four-year service life. Payroll checks can be processed by an outside vendor for $4 each. How many paychecks must be processed per month to justify the purchase of the computer based on a 12% annual interest rate?

HUMAN FACTORS AND SAFETY

Human Factors and Safety Engineering emerged from the disciplines of industrial engineering and psychology. Work is this area is concerned with the effective and safe application of people as elements of a system. The original emphasis was on controls, consoles and cockpits in the military. In recent years, this field has expanded to cover virtually all areas where people interact in engineering systems. Human factors is often called *ergonomics*, especially in European countries.

16.1 Engineering Anthropometry

Engineering anthropometry is the application of scientific measurement techniques to the human body to improve the interaction between people and machines. Many worker performance problems can be eliminated by properly selecting workplace dimensions based on anthropometric measures.

The goal in such studies is to assure that the workplace layout can accommodate the physical capabilities of the majority of people. The typical measure is the *percentile*. A percentile is the percentage of the population having a particular body dimension or capability (such as overhead reach or gripping strength) equal to or less than a particular value. General design limits are usually based on a range from the 5th percentile female to the 95th

Table 16.1 Anthropometric Data for Common Working Positions.

Feature	Percentile Values							
	5th Percentile				95th Percentile			
	Men		Women		Men		Women	
	(in.)	(mm)	(in.)	(mm)	(in.)	(mm)	(in.)	(mm)
Stature - Clothed	66.4	1685	61.8	1568	74.4	1890	70.3	1787
Functional Forward Reach	28.3	726	25.2	640	34.0	864	31.1	790
Overhead Reach, Standing	78.9	2004	73.0	1853	90.8	2305	84.7	2151
Overhead Reach, Sitting	50.3	1279	46.2	1174	57.9	1469	54.9	1394
Functional Leg Length	43.5	1106	39.2	996	50.3	1277	46.7	11186
Kneeling Height	48.0	1219	45.1	1145	53.9	1369	51.3	1303
Kneeling Leg Length	25.2	639	23.3	592	29.7	755	27.8	705

percentile male. For any body dimension, the 5th percentile indicates that 5% of the population will be equal to or smaller than that value. A design range from the 5th percentile of women to the 95th percentile of men should cover a large majority of personnel.

In most applications, the important percentile for the selection of workplace dimensions is the 5th percentile. For example, the 5th percentile in women for overhead reach height is 73 inches (1835 mm). Consequently, a workspace designed for both men and women should not require overhead reaching over 73 inches. Table 16.1 gives anthropometric data for common working positions.

16.2 Lighting and Workplace Effectiveness

The illumination level in a workplace can strongly effect worker performance. Proper lighting is necessary for the safe and efficient execution of tasks.

The quantity of light emitted from a light source in unit time is called *luminous flux*. It is expressed in the units of lumens (lm). *Luminance* is a measure of the intensity of light emitted from a light source per unit area normal to the direction of the light flux. Luminance is expressed in *candela/m²* (*cd/m²*). Another important unit of measurement in lighting is *illuminance*, the part of the total light flux that is incident on a given surface. In standard practice, it is the quantity of light with which a work surface is illuminated. The measure of illuminance is the *footcandle* or ft-C (1 ft-C = 10.8 *lm/m²* or 10.8 lux).

The recommended illumination level of a workplace is determined by the type of task that is performed. Inadequate illumination can result in poor efficiency, fatigue or damage to eyesight. Table 16.2 lists recommended levels of illumination for different types of workplaces and tasks.

16.3 Noise and Vibration

Noise is frequently defined as "unwanted sound." Noise or any sound is measured by frequency in Hertz (Hz) and its intensity in decibels (dB). Noise has several undesirable side effects: continuous exposure to intense noise can cause deafness. Noise also interferes with communication and prevents recognition of warning signals.

Measurements of sound intensity used in human factors engineering are typically made on the A-weighted scale (dBA). This scale is made up of deemphasized frequencies less than 1000 Hz (the ear is less sensitive to these frequencies). Extremely intense noise for prolonged periods or impulses of intense noise (gunfire or explosions) over 120 dBA can cause permanent damage

Table 16.2 Recommended Ranges of Illumination.

Type of Activity or Area	Recommended Illumination	
	ft-C	lux
Public Areas with Dark Surroundings	2 - 5	20 - 50
Areas for Brief Visits	5 - 10	50 - 100
Working Areas where Visual Tasks are Occasionally Performed	10 - 20	100 - 200
Performance of Visual Tasks of High Contrast or Large Size (reading printed matter or rough assembly)	20 - 50	200 - 500
Performance of Visual Tasks of Low Contrast or Small Size (reading handwritten text or difficult inspection)	100 - 200	1000 - 2000
Performance of Visual Tasks of Extremely Low Contrast and Small Size (surgical procedures or circuit board repair)	1000 - 2000	10000 - 20000

to hearing. Industrial noise at levels of 90 dBA to 110 dBA can also cause permanent hearing loss if experienced over a period of months.

Most industrialized countries have legally enforceable maximum noise levels for workers. The Occupational Health and Safety Act of 1970 (OSHA) developed maximum noise exposure standards for all employees. The noise exposures permitted under OSHA are given in Table 16.3.

Personal protective equipment, such as earplugs, earmuffs and helmets should be supplied to employees in areas where noise abatement is difficult or expensive.

Vibration is characterized in terms of its frequency, amplitude, and direction. It can affect health in a variety of ways. One of the most important is the resonance frequency range of the human body. Vibrations in the range of 4 Hz to 8 Hz cause internal organs to resonate. Prolonged exposure to vibrations of approximately 1-g acceleration in the resonant range can cause abdominal pain, loss of equilibrium, nausea, and shortness of breath. Sequential vibrations to the hands in the range of 1.5 g to 80 g with frequencies from 8 Hz to 5000 Hz are

Table 16.3 Noise Exposures Permitted by OSHA.

Duration Per Day (hours)	Sound Level (dBA)
8	90
6	92
4	95
3	97
2	100
1.5	102
1	105
0.5	110
0.25 or less	115

also of concern. Tools may transmit such localized vibrations to the hands. Problems such as stiffness, numbness, pain, and loss of strength may result from prolonged exposure.

16.4 Repetitive Motion

Repeated simple motions during the workday can result in a variety of health problems that can have severe consequences. Repetitive motion disorders or *cumulative trauma disorders* (CTD) can result from the execution of a simple task, such as raising an arm overhead or using a screwdriver repeatedly. Typical motion patterns that can cause risk are: bending of the wrist, grasping or pinching objects, raising an arm overhead or applying a large amount of force with the hand.

Repetitive motion injuries occur over months of executing the same motion pattern without an opportunity for the body to recover. Pain and minimized movement can result from a variety of repetitive motion related ailments. A repeated motion in an awkward position or requiring a high application of force coupled with a lack of rest cause these types of injuries. Some common forms of repetitive motion problems are:

• **Tendinitis** Inflamed and sore tendons causing pain, swelling and weakness.

• **Carpal Tunnel Syndrome** Excessive pressure on the median nerve in the wrist causing numbness, tingling and pain in the wrist.

• **Rotator Cuff Injury** Inflamed tendons in the shoulder causing pain and limited motion.

• **Tenosynovitis** Swelling of the tendon and the sheath that covers it causing tenderness and pain.

Repeated motion injuries can be prevented by proper tool selection, minimized force application requirements and variety of tasks being performed (providing an opportunity for the body to recover).

Practice Problems

16.1   What is the minimum standing height that should be designed in a walk-in freezer?

16.2   A service counter is being designed. How far should employees be expected to reach across the counter?

16.3   What is the minimum illumination that should be provided in a walk-in freezer?

16.4   What is the minimum recommended illumination for a truck loading dock?

16.5   What is the maximum sound level that can be present in a typical factory?

16.6   A truck seat resonates at 5 hz. What is the potential problem with this design?

16.7   What is the risk associated with using a screwdriver throughout the workday?

Part 6

Manufacturing Planning and Strategy

ENGINEERING MATERIALS

17.1 Fatigue Properties

If a material is subjected to fluctuation or to a number of cycles of stress reversal, failure may occur. This failure may occur though maximum stress at any cycle is less than the value at which failure would occur under constant stress. By subjecting test specimens to stress cycles and, in turn, counting the number of cycles to failure, fatigue properties may be determined.

So if a metal is stressed repeatedly, and finally breaks, the break was caused by fatigue.

With a series of these tests and when the maximum stress values are reduced in a progressive manner, *S-N* diagrams can be plotted. The *S* (stress) is on the vertical axis and the *N* (number of cycles to failure) is on the horizontal axis. A sample *S-N* diagram is shown in Figure 17.1. For a material without an endurance limit, standard practice is to specify fatigue strength as a stress value. This stress value corresponds to the number of stress reversals.

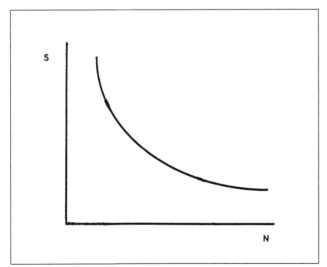

Figure 17.1 Typical Stress-to-cycles Comparison.

17.2 Stress and Strain

When an external force is applied to a body which changes its size or shape, the body resists this external force. The internal resistance of the body is known as *stress* and the accompanying changes in dimensions of the body are called *deformations* or *strains*. The total stress is the total internal resistance acting on a section of the body. An increase in length of the body created by a force results in tensile strain (+); a decrease is compressive strain (-); and an angular distortion is a shear strain.

The *tensile test* as specified by the American Society for Testing and Materials (ASTM) determines mechanical properties of materials in relation to stress and strain. Next to the hardness test, the tensile test is the most frequently used.

A tensile test will reveal several mechanical properties which play a major role in engineering design: proportion limit, elastic limit, yield point, yield strength, ultimate strength, breaking (rupture) strength, ductility, and modulus of elasticity.

The following definitions can be referenced in Figure 17.2 and Figure 17.3.

Proportional Limit. An early part of the stress-strain graph may be approximated by a straight line *OP* in both Figures 17.2 and 17.3. In this range, the stress and strain are proportional, so any increase in stress results in a proportionate increase in strain. The stress at the limit of proportionality (point *P*) is known as the *proportional limit*.

Elastic Limit. If a small load is applied to a material and then removed, that material will indicate a zero deformation. This means the strain is elastic. If the load is increased enough, a point will be reached where the material experiences a permanent deformation. This is called the *elastic limit*. Therefore, the definition of elastic limit is the

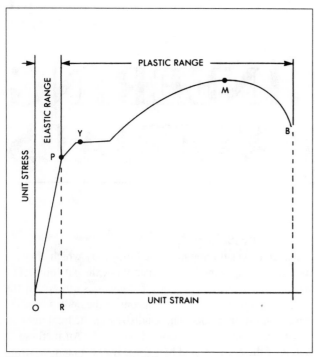

Figure 17.2 Stress-strain Diagram for Ductile Steel.

minimum stress at which permanent deformation occurs. For most materials, the numerical value of elastic limit and proportional limit are nearly the same.

Yield Point. Point *Y* in Figure 17.2 represents the yield point of a ductile material. The *yield point* can be defined as the point at which permanent deformation continues without an increase in stress. In some ductile materials, the stress (load) may actually decrease momentarily, resulting in an upper and lower yield point.

Yield Strength. Most nonferrous materials and the high-strength steels do not possess a well-defined yield point. Therefore, the yield strength must be determined by the offset method. The *yield strength* is the stress at which a material exhibits a specified limiting deviation from the proportionality of stress to strain. Figure 17.3 exemplifies the offset method. *OX* is laid off along the strain axis, then a line *XW* is drawn parallel to *OP*, thus locating *Y* (yield strength). The intersection of the offset is generally between 0.1% and 0.2% of the gage length (extensometer, 2 inches (50.8 mm)).

Ultimate Strength (tensile strength). As the load on the test piece is increased still further, the stress and strain increase, as indicated by the portion of the curve *YM* in Figure 17-2 for a ductile material, until the maximum stress is reached at point *M*. The ultimate strength or tensile strength is, therefore, the maximum stress developed by the

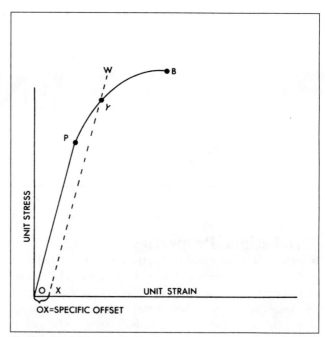

Figure 17.3 Stress-strain Diagram for Brittle Material.

material, based on the original cross-sectional area. A brittle material breaks when stressed to the ultimate strength (point *B* in Figure 17.3), whereas a ductile material will continue to stretch. Tensile strength is measured in psi.

Breaking Strength. The *breaking* or *rupture strength* is that point of the curve where the specimen actually fails. At maximum stress, ductile materials experience a localized deformation. As the cross-sectional area decreases at a rapid rate, the stress decreases. The deformation and elongation occur rapidly until failure. With a brittle material, the ultimate strength and breaking strength coincide.

Modulus of Elasticity. The modulus of elasticity (*E*) is an indication of the stiffness of a material. The formula for computing *E* is:

$$E = \frac{\text{Stress}}{\text{Strain}} = \frac{\gamma}{e}$$

It is important to remember that the equation for computing *E* represents the slope of the curve on the stress-strain diagram in the elastic region. Therefore, if a material is heat treated to increase the yield and ultimate strength, the value of *E* remains unchanged.

There are three important properties of metal that must be addressed in design engineering.

The first, ductility, was discussed previously. *Ductility*, defined as the amount of deformation a material can withstand until failure, is determined by the tensile test. It is measured by determining the percent of elongation at fracture.

The second—and most widely used—is the hardness properties of a material. *Hardness* is that property of a material that enables it to resist plastic deformation. There are three basic types of hardness tests: Brinell, Rockwell, and Vickers. The two most commonly used in industry are the Brinell and the Rockwell.

Brinell Hardness Test. The *Brinell hardness number* (BHN) is the ratio of the load in kilograms to the impressed area in square millimeters. The Brinell test consists of pressing a 10-mm (0.4-inch) diameter ball into a material under a specified load for a specific amount of time. (For nonferrous materials, the load is 500 kg (1100 lb) for 10 seconds; ferrous materials require 3000 kg (6600 lb) for 10 seconds.) The diameter of the impression is measured by means of a microscope containing a scaled ocular, graduated in tenths of a millimeter, permitting estimates to the nearest 0.01 mm (0.0004 inch). The following formula is used to calculate BHN:

where: L = Test Load, kg
(D) = Diameter of ball, mm
d = diameter of impression, mm

then:

$$BHN = \frac{L}{(\pi\,D/2)(D-D^2-d^2)}$$

Rockwell Hardness Test. The *Rockwell test* uses a direct-reading instrument based on the principle of differential depth measurement. There are two basic types of Rockwell machines: the normal tester for relatively thick sections and the superficial tester for thin sections. The Rockwell machine operates by placing a minor load on a specimen (10 kg (22 lb) for normal and 3 kg (6.6 lb) for superficial) and zeroing the gage. The major load is then applied. After the gage comes to a rest, the major load is relieved and the reading taken. The major load is usually 60 kg (132 lb), 100 kg (220 lb), and 150 kg (330 lb) for normal testing and 15 kg (33 lb), 30 kg (66 lb), and 45 kg (99 lb) for superficial. The penetrators consist of steel balls of various sizes and a 120° conical diamond. The most commonly used Rockwell scales are B (100 kg (220 lb) major load, 1.6-mm (0.0625-inch) ball) and C (150-kg (330-lb) major load, diamond).

The third property of metal to be considered is that of toughness. *Toughness* is the ability of a material to absorb energy from impact. Toughness can be derived from stress-strain graphs such as those depicted in Figure 17.2 and Figure 17.3. Material in Figure 17.2 is considered to be tougher than the material in Figure 17.3. Toughness is mainly a property of the plastic range since only a small part of the total energy absorbed is elastic energy that can be recovered when the stress is released.

Toughness can also be determined by an impact test. This method uses a machine with a swinging pendulum of fixed weight raised to a standard height. The principle is that the pendulum swings with a definite kinetic energy; upon striking and breaking the specimen, the amount of energy used is recorded in foot-pounds. It is a comparison used in a basic toughness determination.

Cold Treating. Cold treatment of steel consists of exposing the metal parts on tools to subzero temperatures to obtain the desired conditions or properties. Such treatment can provide improved strength, dimensional or microstructural stability, greater resistance to wear, stress relieving, and retarded aging. Low ductility is also associated with the cold working of steel.

For steels, proper cold treatment ensures a more uniform and completely transformed microstructure. Previously, soft, retained austenite was transformed into hard, more stable martensite, which could be subsequently tempered.

17.3 **Heat-treatment of Metals**

Heat-treatment may be defined as any process whereby metals are better adapted to desired conditions or properties, in predictably varying degrees, by means of controlled heating and cooling in their solid state without altering their chemical composition. Heat-treatment can be applied to a variety of metals: iron, steel, aluminum, copper, and numerous others. However, because of the versatility and broad use of ferrous alloys in industry, the treatments applied to steel are, by far, the most widely used.

Heat-treatment may be employed to improve tensile strength, ductility, toughness, wear resistance, machinability, formability, bending quality, corrosion resistance, magnetic properties, and other properties. Heat-treatment temperatures range from room temperature to 2350°F (1287°C) and as low as -325°F (-198°C).

Full Annealing. *Full annealing* consists of heating the steel to the proper temperature and then cooling slowly through the transformation range, preferably in the furnace or in any good heat-insulating chamber. The purpose of annealing may be to refine the grain, induce softness, improve electrical and magnetic properties and, in some cases, improve machinability.

The proper annealing temperature for a hypoeutectoid (low-carbon) steel is 50°F (10°C) above the upper-critical-temperature (A_c3) line shown in Figure 17.4. Refinement of the grain size of hypereutectoid (high-carbon) steel will occur about 50°F (10°C) above the lower-critical-temperature $(A_{c3,1})$ line. Heating above this temperature will coarsen the austenic grains, which, on cooling, will transform to large pearlitic areas surrounded by a network of cementite. This creates a plane of weakness, brittleness, and usually poor machinability. Annealing never should be a final heat treatment for hypereutectoid steels.

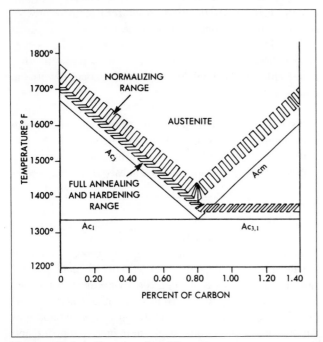

Figure 17.4 A Heat-treating Diagram.

Spheroidize Annealing. This is a process sometimes used to improve the properties of hypereutectoid steels. It involves holding the material for a prolonged period at a temperature just below the lower critical temperature or heating and cooling alternately between temperatures slightly above and slightly below the lower critical line. Spheroidize annealing allows cementite to assume the form of round particles (spheroids) instead of pearlitic plates. This structure not only gives good machinability but also high ductility.

Stress-relief Annealing. Also referred to as *subcritical annealing*, this process is useful in removing residual stresses due to heavy machining or other cold-working processes and welding. It is usually carried out at temper-

atures below the lower critical temperature (1000°F (538°C) to 1200°F (649°C)).

Normalizing. This process is carried out by heating the material approximately 100°F (38°C) above the upper-critical-temperature (Ac3 or Acm) line, followed by cooling in still air to room temperature. The purpose of normalizing is to produce a steel harder and stronger than full annealing, refine the grain, homogenize the structure, and improve machinability, particularly in low-carbon (hypoeutectoid) steels. Cooling in air causes the austenite to transform into pearlite at a lower temperature, resulting in a pearlite structure finer than that of annealing. Figure 17.4 shows a relationship between annealing and normalizing.

Hardening. Although hardening increases the natural attributes of steel, it creates a structure of martensite, a supersaturated solid solution of carbon and iron, which is extremely hard and brittle. Formed by rapid cooling from the upper critical temperature or austenite range to room temperature, martensite appears as a needle-like structure when viewed under a microscope. The carbon atoms are trapped in a solid solution of iron before they are able to diffuse out of solution. The resultant structure is body-centered-tetragonal, which is a highly stressed condition. It is important that the surface of the material be clean and completely free of nicks, pits, seams, or tool marks, because these imperfections can become stress risers during the hardening process.

Hardenability—the ability of a material to become hard—is influenced greatly by the composition of that material. The higher the carbon content, the greater the hardenability; thus the higher the hardenability, the less severe the quenching medium must be to obtain the same hardness. The Jominy end quench test determines a material's hardenability.

This list of quenching mediums is listed in increasing order of severity:

1. Gases (air),
2. Oil,
3. Water, and
4. Brine.

17.4 Surface Hardening
Surface-hardening treatments are used to add carbon, nitrogen, or both to the surface of steel parts to provide a hardened layer or case of a definite depth. Sometimes called *case hardening*, these treatments create a surface with high wear resistance but maintain a ductile inner core for impact strength or toughness.

There are five principal methods of case hardening.

Carburizing. The oldest and one of the least expensive methods of case hardening, carburizing consists of placing a low-carbon steel, usually about 0.2% carbon or lower, in an atmosphere containing substantial amounts of carbon monoxide. The usual carburizing temperature is 1700°F (927°C). At that temperature, carbon is dissolved into the austenite.

Commercial carburizing may be accomplished by pack carburizing, gas carburizing, or liquid carburizing. In pack carburizing, the work is surrounded by a carburizing compound (charcoal, etc.) in a closed container. With gas carburizing, the work is placed in contact with gases rich in carbon monoxide and hydrocarbons, such as methane, butane, and propane. Liquid carburizing is performed in a bath of molten salt containing up to 20% sodium cyanide. The cyanide is a source of carbon and nitrogen. The case obtained by this method is composed largely of carbon with only a small amount of nitrogen.

Cyaniding. Cyaniding consists of immersing the steel in a molten bath containing about 30% sodium cyanide at temperatures between 1450°F (788°C) and 1600°F (871°C) and usually followed by water quenching. Cyaniding differs from liquid carburizing in the composition and character of the case. The process creates a high-nitrogen and low-carbon case, quite the reverse of liquid carburizing. Used mainly where a light case is required, cyaniding requires only one hour for a 0.01-inch- (0.3-mm) deep case.

Nitriding. In nitriding, the part is placed in an airtight container through which ammonia is passed continuously at a temperature between 900°F (482°C) and 1150°F (621°C). Under these conditions, the ammonia partially decomposes into nitrogen and hydrogen. The nitrogen penetrates the steel surface and combines with iron-forming nitrides. Nitriding produces the hardest surface of the case-hardening processes; however, it is the most time-consuming process (50 hours = approximately 0.015-inch (0.381-mm) deep).

Flame Hardening This process involves rapidly heating a selected surface area of medium- or high-carbon steel and immediately cooling it in water or by airblast.

Induction Hardening. Induction hardening is similar to flame hardening. They are both shallow hardening and do not change the chemical composition of the steel. It is used mainly on medium-carbon steels. In this method, the part to be hardened is made the secondary of a high-frequency induction apparatus. The primary, or work, coil consists of several turns of water-cooled copper tubing. When a high-frequency alternating current is passed through the coil, a magnetic field is set up, inducing a high-frequency eddy current in the metal. The losses due to these currents produce the required heat steels.

17.5 Plastics

Plastics are nonmetallic materials that can be formed and shaped by many methods. They can be made from such natural resins as shellac; however, most plastics used in industrial applications are produced from man-made synthetic resins.

Plastics became one of the most common classes of engineering materials in the past decade. For the last several years, the production of plastics, on a volume basis, exceeded steel output.

To ensure selection of the most suitable plastics material from among the hundreds available, it is advisable to develop direct sources of information, including contacts with resin manufacturers. Table 17-1 lists representative properties for selected thermoplastics and thermosets commonly used in industrial applications.

ASTM Classification System. The American Society of Testing and Materials (ASTM) is a technical organization bringing manufacturers, specifiers, and users together to standardize specifications and test methods. Several individual types of plastics have been covered by specific ASTM standards. Standards are useful for two reasons: first, the specifier can be assured of minimum strengths and properties for design calculations and, second, competitive manufacturers' resins can be used. Unfortunately, the rapid growth of the plastics industry also limits the usefulness of single standards. As mentioned, considerable competition exists between different types of resins.

Basic Terminology. The term "polymer" is commonly used interchangeably with the term "plastics," yet neither is entirely accurate in its delineation. Plastic means pliable, yet most engineering polymers are not plastic at room temperature. Polymer, on the other hand, can include every kind of material made by polymerization with repeating molecules. The ASTM definition of a plastic is: "A material that contains, as an essential ingredient, an organic substance of large molecular weight, is solid in its finished state, and, at some stage in its manufacture or in its processing into finished articles, can be shaped by flow."

In broad terms, plastics are man-made polymers. Polymer is the generic name for materials composed of long,

TABLE 17-1
Properties of Selected Industrial Plastics

Type of Plastics	Molecular Packing	Specific Gravity	Mechanical Properties (Room Temperature)			
			ASTM D-638 Tensile Strength, psi (MPa)	ASTM D-638 Elongation, percent	ASTM D-695 Compressive Strength, psi (MPa)	ASTM D-256 Impact Strength (Izod), ft-lb/in. (J/cm)
Polystyrene	Amorphous	1.10	7500 (51.7)	2	14,000 (96.5)	0.3 (0.2)
High-impact polystyrene	Amorphous	1.15	5000 (34.5)	10	7500 (51.7)	0.6-10.0 (0.3-5.3)
Acrylics	Amorphous	1.15	10,000 (69.0)	6	15,000 (103.4)	0.4 (0.2)
Polycarbonate	Amorphous	1.20	9000 (62.1)	100	10,000 (69.0)	15.0 (8.0)
ABS	Amorphous	1.05	6000 (41.4)	30	8000 (55.2)	6.0 (3.2)
Acetal (homopolymer)	Crystalline	1.40	10,000 (69.0)	40	18,000 (124.1)	1.8 (1.0)
Nylon 6/6 at 50% RH*	Crystalline	1.15	11,000 (75.8)	400	10,000 (69.0)	2.1 (1.1)
Polypropylene	Crystalline	0.91	4500 (31.0)	500	7000 (48.3)	1.0 (0.5)
Polyethylene (high-density)	Crystalline	0.95	4000 (27.6)	600	3000 (20.7)	10.0 (5.3)
Polyet hylene (medium-density)	Crystalline with amorphous regions	0.93	2400 (16.5)	600	3000 (20.7)	8.0 (4.3)
Polyethylene (low-density)	Semicrystalline	0.91	1500 (10.3)	700	3000 (20.7)	No break
Epoxy	Cross-linked network	1.25	10,000 (69.0)	3	20,000 (137.9)	0.8 (0.4)
Phenolic	Cross-linked network	1.35	7000 (4-8.3)	2	10,000 (69.0)	0.4 (0.2)

*RH = relative humidity

chain-like molecules. Most living tissue and cells are polymeric. Plastics are created either by modifying natural polymers, such as cellulose fibers, or by causing small synthetic molecules to bond together into a chain. Compared to other classes of materials, the plastics molecular chain is enormous, giving it the term "macromolecule." Millions of macromolecular chains must be put together to make industrially useful quantities.

Mechanical Properties. Both metals and plastics are characterized by similar types of mechanical properties. Metals, however, tend to be consistent in the sense that their behavior is adequately characterized by stress-strain relationships. In contrast, while the individual plastics materials also display distinctive stress-strain characteristics, the mechanical properties of plastics are more dependent on the additional factors of temperature and time (under load). In the design application, and to some extent in processing, creep data are of significant importance in the field of plastics materials.

The engineering plastics materials have ultimate tensile and compressive strengths and stiffness properties significantly lower than those of metals. This difference is especially meaningful when comparing plastics to tool steels and high-strength steels. However, the differential in ultimate mechanical strength is much less when plastics are compared with metals such as aluminum, magnesium, zinc, and copper.

Thermoplastic engineering resins are characterized as those resins with the following combination of properties:

• Thermal, mechanical, chemical, corrosion resistance, and fabricability;

• Ability to sustain high mechanical loads, in harsh environments, for long periods of time;

• Predictable, reliable performance.

Processing and Applications. The melt-processability of thermoplastic resins is a basic characteristic that distinguishes them from thermosets. This pertains not only to the advantages of injection molding as compared to compression or transfer molding, but also to the variety of processing alternatives that expand the utility of thermoplastics. There are thermosets that can be injection molded, but only the thermoplastics offer the options of blow molding or extrusion into sheet, film, and profiles.

The degree to which each of the engineering plastics is amenable to alternative processing methods varies. The relative potential of each depends also on its potential in alternative processes, not just on its utility in injection molding. Tables 17-2 and 17-3 present a general relationship between thermosets and thermoplastics and their respective parts production processes.

Nylon, in addition to its use in injection molding, can be extruded into monofilament and brush filament. Nylon-6 is used for sewing thread, fishing line, household/industrial brushes, and level-filament paint brushes. Nylon-6,6, stiffer than nylon-6, is used for sewing thread and household/industrial brushes. Nylon-6,12 dominates in personal-care brushes, although polyethylene terephthalate (PET) competes in these applications. In tapered-filament paint brushes, nylon-6,12's leading position has been taken over by PBT, a more expensive but more versatile filament.

Nylon is used as a wire coating, primarily as a protective abrasion-resistant coating over PVC-insulated wire. Nylon film can be cast, or blown, or extrusion-coated onto various substrates. Most nylon film is cast, and virtually all is sold to converters who add a sealant layer of low-density polyethylene (LDPE), ethylene-vinyl acetate copolymer (EVA), or ionomer. It is mainly used for vacuum packages of processed meats and cheese, usually combined with a PET-sealant cover web. Nylon film is also used for fresh-meat packaging, and a new market has opened in medical device packaging using techniques similar to those for formed-meat packaging. The most important properties in these applications are formability and heat resistance.

Nylon strapping began replacing steel strapping in the early 1960s, even at higher cost, because of the general advantages of nonmetallic strapping. In recent years, nylon met increasing competition in this market from polypropylene and PET.

Nylon is also extruded into rods, tubes, and shapes for machining, an important option for low-volume runs. The blow molding of nylon has been restrained partly by cost and partly by the difficulties inherent in crystalline resins because of their sharp melting point. Nylon blow-molding resins, developed with high melt strengths for parison forming, are used to some extent for monolayer and coextruded bottles and for gas tanks in small equipment. Nylon-6 is also cast to produce very large bearings.

Nylon-11 is used for powder coatings and for flexible tubing. Nylon-12 is used for the same purposes, but to a greater extent in Europe than in the U.S. These resins have exceptional moisture resistance, but they are considerably less stiff than nylon-6 or -6,6. They are used to some extent in rotational molding.

In contrast to nylon, acetal offers few options outside the injection molding category. An acetal terpolymer is available for injection blow molding, but, apart from some carburetor floats and rod extrusions, it has found little usage. Although acetal is difficult to extrude, it is extruded into shapes, as is nylon, for subsequent machining. Almost all acetal consumption is in injection molding, a factor that limits its total consumption.

The PET thermoplastic polyesters used for film, sheet, and blow molding are not the same as those used for injection-molded engineering applications. PBT can be blow molded but rarely is. While used almost entirely in injection molding, PBT is used in tapered brush filaments and in extruded strip for small electrical parts.

TABLE 17-2
Thermoplastics Parts Manufacturing Processes

Thermoplastics	Compression/molding	Transfer molding	Injection molding	Extrusion	Rotational molding	Blow molding	Thermoforming	Reaction injection molding	Casting	Forging	Foam molding	Reinforced plastic molding	Vacuum molding	Pultrusion	Calendering
Acetal			●	●	●	●	●				●	●			●
ABS			●	●	●	●	●			●					●
Acrylic	●		●	●		●	●		●						●
Cellulose acetate	●		●	●			●								●
Cellulose acetatebutyrate	●		●	●	●		●								●
Cellulose nitrate	●		●	●			●								●
Cellulose propionate	●		●	●			●								●
Ethyl cellulose			●	●			●					●			●
Chlorinated polyether			●	●	●	●	●								
CTFE	●	●	●	●			●								●
Tetra-fluoroethylene (TFE)	●	●	●	●											●
FEP	●	●	●	●		●									●
CTFE-VF$_2$	●	●	●	●		●									●
Nylon			●	●	●	●		●	●	●		●			●
Phenoxy			●	●		●									
Polyimide			●												
Polycarbonate			●	●		●	●								●
Polyethylene			●	●	●	●	●			●	●	●			●
Polyphenylene oxide (PPO)			●	●		●						●			
Polypropylene (PP)	●		●	●	●	●	●		●	●		●			●
Polystyrene			●	●	●	●	●				●	●			●
Polysulfone			●	●		●	●					●		●	
Polyurethane			●	●	●						●	●	●		●
SAN			●	●		●									
PVC	●	●	●	●	●	●	●				●	●	●		●
Polyvinyl acetate	●	●	●	●	●	●	●				●				●
Polyvinylidene chloride			●												

TABLE 17-3
Thermoset Plastics Parts Manufacturing Processes

Thermosetting Plastics	Compression/ molding	Transfer/ molding	Injection molding	Rotational molding	Thermoforming	Reaction injection molding	Casting	Foam molding	Reinforced plastics molding	Laminating
Alkyd	●	●	●				●		●	
Allyd					●		●		●	●
Epoxy				●		●	●	●	●	●
Melamine	●	●	●					●	●	●
Phenolic	●	●	●				●	●	●	●
Polyester (unsaturated)	●					●	●	●	●	
Polyurethane						●				
Silicone							●	●	●	
Urea	●	●	●					●		

Noryl resin's use in extrusion is relatively minor compared to injection molding, but it is used to some extent for stock shapes and to an increasing extent for sheet and profiles. Noryl (General Electric's PPO) sheet competes with flame-retardant ABS as it does in injection molding, and it can compete with less expensive resins like ABS and PVC, where its properties permit the extrusion of thinner walls.

The transparency of polycarbonate, combined with its extrudability and impact resistance, makes it a strong competitor for acrylic sheet in replacing flat glass. Extruded sheet for glazing, lighting, and signs accounts for approximately 25% of polycarbonate's volume. Its use in extruded profiles is minor, but polycarbonate is widely used in blow molding for water bottles, milk bottles, baby nursing bottles, and miscellaneous packaging.

Thermoset Plastics Molding. Types of thermoset materials capable of being molded include phenolic, urea, melamine, melamine-phenolic, diallyl phthalate, alkyd, polyester, epoxy, and the silicones. Thermosetting molding compounds processed from the individual heat-reactive resin systems are available in a wide range of formulations to satisfy specific end-use requirements. Depending upon the type of material, products may be supplied in granular, nodular, flaked, diced, or pelletized form. Polyester materials are supplied in granular, bulk, log, rope, or sheet form, and polyurethanes are made in many forms, ranging from flexible and rigid foams to rigid solids and abrasion-resistant coatings.

As the term implies, thermoset molding compounds, when placed within the confines of a mold (generally hardened steel), are subjected first to heat to plasticize and cure the material then to pressure to form the desired shape. The mold is held closed, under pressure, sufficiently long to polymerize or cure the material into a hard, infusible mass.

Injection Molding. Injection molding is a versatile process for forming thermoplastic and thermoset materials into molded products of intricate shapes, at high production rates, and with good dimensional accuracy. Injection molding makes use of the heat-softening characteristics of thermoplastic materials. These materials soften when heated and reharden when cooled. No chemical changes take place when the material is heated or cooled; the change is entirely physical, allowing the softening and rehardening cycle to be repeated several times. While this is true for thermoplastics, a chemical reaction does occur with certain thermosets and rubbers that can be injection molded.

The basic injection molding process involves injecting, under high pressure, a metered quantity of heated and plasticized material into a relatively cool mold-in, which is solidified by the plastics material.

Extrusion Forming. The extrusion process is a continuous operation in which hot plasticized material is forced through a die opening that produces an extrudate of the desired shape. The most commonly extruded materials are rigid and flexible vinyl, ABS, polystyrene, polypropylene, and polyethylene. Nylon, polycarbonate, polysulfone, acetal, and polyphenylene are included among other plastics that can be extruded.

The extrusion process is used to produce film (thinner than 0.030 inch (0.76 mm)), sheets (thicker than 0.030 inch (0.76 mm)), filaments, tubes, and a variety of profiles. The process of plastics extrusion also is used to coat cables, wires, and metal strips.

Reaction Injection Molding. Reaction injection molding (RIM) is a form of injection molding that brings temperature and ratio-controlled, liquid reactant streams together under high-pressure impingement mixing to form a polymer directly in the mold. Two liquid reactants (monomers) are mixed together as they enter the mold. A chemical reaction produces the plastics as it forms the part.

When compared to other molding systems, RIM offers more design flexibility, lower energy requirements, lower pressures, lower tooling costs, and lower capital investment. Significant advantages in design and production are gained from the RIM fabricating capability for incorporating a load-bearing, structural skin and a lightweight, rigid, cellular core into a part in one processing operation.

Reinforced Thermoset Plastics. The reinforced plastics described in this section primarily encompass polyester and fiberglass systems. The most commonly used processes can be divided into two categories: high- and low-volume. Several less common intermediate volume processes also are described.

The low-volume processes fall in the 2000 to 25,000 parts-per-year range. They are characterized by essentially low-pressure to pressureless hand or spray lay-up in low-cost molds, with a high labor cost. High-volume processes are those in which more than 30,000 parts per year are produced (100,000 parts per tool for automotive components). They involve an initial high cost for tooling and equipment, but the labor intensity is low. These processes are not competitive with the metal stamping process unless they eliminate the need for multipiece assembly operations.

The high-volume processes for reinforced thermoset plastics are sheet molding compound (SMC), thick molding compound (TMC), bulk molding compound (BMC), pultrusion, and pulforming.

Thermoforming Plastic Sheet and Film. Thermoforming consists of heating a thermoplastic sheet to its processing temperature and forcing the hot, flexible material against the contours of a mold. This pliable material is rapidly moved either mechanically with tools, plugs, matched molds etc., or pneumatically with differentials in pressure created by a vacuum or by compressed air.

Thermoforming has several advantages: (1) low costs for machinery and tooling because of low processing pressures; (2) low internal stresses and good physical properties in finished parts; (3) capability of being predecorated, laminated, or coextruded to obtain different finishes, properties, etc.; (4) capability of forming light, thin, and strong parts for packaging and other uses; and (5) capability of making large, one-piece parts with relatively inexpensive machinery and tooling. The main disadvantages are: (1) higher cost of using sheet or film instead of plastics pellets and (2) necessity of trimming the finished part.

Numerous special techniques are used in processing plastics in addition to the basic procedures described. Combinations of several processes often may be used advantageously for producing specialty plastics products.

Blow Molding. Blow molding is a process for shaping thermoplastic materials into one-piece, hollow articles by means of heat and air pressure. The method consists basically of stretching a hot thermoplastic tube with air pressure, then hardening it against a relatively cool mold. A wide variety of blow-molding techniques and equipment is used to suit specific applications.

The principal difference between blow molding glass and blow molding plastics stems from the difference in the properties of the materials. Molten glass is much less sensitive than plastics, not only in melt viscosity, but also in chemical stability. For this reason, modified glass-blowing machines cannot be used for plastics. Another difference is that glass blowers start with a drop of molten glass, while the plastics blowers use a preformed plastics part, usually a tube of molten plastics called a parison.

Liquid Injection Molding. In comparison to other processes, the liquid injection molding (LIM) method has the potential to replace compression and transfer molding of thermoset plastics in some applications. Instead of being charged into the cavity of a compression mold or a transfer pot as powder, pellets, or other molding compound, two LIM material components are pumped directly from the shipping containers through a mixing device and injected into a heated mold. Here, they cure.

Liquid injection molding differs considerably from reaction injection molding (RIM). The pumping systems used in LIM operate at generally lower pressure than those of RIM, and mixing in LIM more often is accomplished mechanically than by impingement. Also, while RIM parts usually are large, LIM parts typically are quite small.

Rotational Molding. Rotational molding is a process for forming hollow plastics parts. The process uses the principle that finely divided plastics material becomes molten in contact with a hot metal surface and then takes the shape of that surface. When the polymer is cooled while in contact with the metal, a reproduction of the mold's interior surface is produced.

Structural Foam Molding. A structural foam is a plastics product with a rigid cellular core and an integral skin. The solid skin is typically 0.030 to 0.080 inch (0.76 to 2.03 mm) thick. Density reductions compared with solid plastics are in the range of 20% to 40%, depending upon part configuration, thickness, and molding conditions. Structural foams can be molded and extruded. Since the bending stiffness of parts increases in proportion to the cube of the part's thickness, structural foam parts can be made quite rigid, with good strength-to-weight ratios. All thermoplastics can be foam-molded.

Casting. Casting processes are applicable to some thermoplastics and thermosets. These materials can be cast at atmospheric pressure in inexpensive molds to form large parts with section thicknesses that would be impracticable for other manufacturing processes. Casting resins are molded on a production basis in lead, plaster, rubber, and glass molds.

Forging. The forging process sometimes is used in manufacturing thermoplastic parts that would be difficult to produce by other processes. The forging technique is capable of producing thick parts with abrupt changes in sections. One use for forging is in producing thick, large-diameter gears from polypropylene.

Machining. For low-volume production or prototypes, most plastics can be machined with standard woodworking and metalworking equipment and cutting tools. However, different tool angles are necessary when high-volume parts production is required. When more than a few pieces are to be machined, the need for cooling should be given consideration.

Although most plastics can be machined dry, some require a coolant, particularly when machined at high speeds and feeds. Heat generated by the cutting friction of the plastics and the metal cutting tool is mostly absorbed by the cutting tool. Therefore, very little heat is transferred to the core of the material. This heat must be either kept to a minimum or removed by a coolant for optimum results.

17.6 Advanced Composites

A composite material is created by the combination of two or more materials—a reinforcing element and a compatible resin binder (matrix) to obtain specific characteristics and properties. The components do not dissolve completely into each other or otherwise chemically merge, although they do act synergistically. Normally, the separate components can be physically identified, as well as the interface between components.

A common type of a composite material is fiberglass. Glass fibers, though very strong, fracture readily if notched. And if put in compression, they buckle easily. But by encapsulating the glass fibers in a resin matrix, they are protected from damage; at the same time, the resin matrix transfers applied loads to the unified fibers so that their stiffness and strength can be fully used in tension and compression.

The more advanced structural composites use fibers of glass, carbon/graphite, boron, Kevlar (aramid), and other organic materials. These fibers are very stiff and strong, yet lightweight. The strengthening effects of the fiber reinforcements in composites are derived from (a) the percentage of fibers (fiber-resin ratio), (b) the type of fibers, and (c) the fiber orientation with respect to the direction of the loads.

While advanced fiber, resin-matrix composites are classified as reinforced thermosets, a special technology sets them somewhat apart from other reinforced thermosets. Called "advanced composites," resin-matrix composites can include hybrids, mixtures of fibers in various forms in the resin (usually epoxy) matrix.

To engineers in the field, an advanced composite denotes a resin-matrix material reinforced with high-strength, high-modulus fibers of carbon, aramid, or boron, and is usually fabricated in layers to form an engineered component. More specifically, the term is applied largely to epoxy-resin matrix materials reinforced with oriented, continuous fibers of carbon and fabricated in a multilayer form to make extremely rigid, strong structures. Another characteristic distinguishing composites from reinforced plastics is the fiber-to-resin ratio. This ratio is generally greater than 50% fiber by weight. However, the ratio is sometimes indicated by volume since the weight and volume in composites are similar.

The Matrix. The matrix serves two important functions in a composite: (1) it holds the fibers in place, and (2) under an applied force, it deforms and distributes the stress to the high-modulus fibrous constituent. The matrix material for a structural fiber composite must have a greater elongation at break than the fibers for maximum efficiency. Also, the matrix must transmit the force to the fibers and change shape as required to accomplish this, placing the majority of the load on the fibers. Furthermore, during processing, the matrix should encapsulate the fibrous phase with minimum shrinkage, placing an internal strain on the fibers. Other properties of the composite, such as chemical, thermal, electrical, and corrosion resistance, also are influenced by the type of matrix used.

The two main classes of polymer resin matrices are thermoset and thermoplastic. The principal thermosets are epoxy, phenolic, bismaleimide, and polyimide. Thermoplastic matrices are many and varied, including nylon (polyamide), polysulfone, polyphenylene sulfide, and polyether etherketone. The matrix material must be carefully matched for compatibility with the fiber material and for application requirements. The selection process should cover factors such as thermal stability, impact strength, environmental resistance, processability, and surface treatment of the reinforcing fibers (sizing).

Fiber Types. The unique fiber geometry provides many advantages in an advanced composite. In their fiber form, materials such as carbon/graphite and boron (also known as polycrystalline ceramic fibers) show near-perfect crystalline structure. Parallel alignment of these crystals along the filament axis provides the superior strengths and stiffnesses characterizing advanced composites. Various production methods are used for the different fiber types.

Composite Construction. Composites can be divided into laminates and sandwiches. Laminates are composite materials consisting of layers bonded together. Sandwiches are multiple-layer structural materials containing a low-density core between thin faces (skins) of composite materials. In some applications, particularly in the field of advanced structural composites, the constituents (individual layers) may themselves be composites (usually of the fiber-matrix type).

Applications. By tailoring the materials and fabrication methods, and by modifying structural designs to accommodate their unique properties, advanced composites can be used for applications requiring high strength, high stiffness, or low thermal conductivity. The beneficial feature of many aerospace uses is that the new materials weigh less than the metallic materials they replace.

Applications Overview. Advanced composites containing such materials as carbon/graphite or aramid fibers in an organic resin matrix are currently used mainly by the aerospace industries. However, these stiff, strong, lightweight materials are also used in various other commercial and industrial applications, ranging from aircraft structures to automobiles and trucks, from spacecraft to printed circuit boards, and from prosthetic devices to sports equipment. Products run the gamut from boat hulls and hockey shin guards to an advanced composites hinge for the retractable arm of the space shuttle.

Carbon/graphite cloth reinforced plastics are used in a variety of applications requiring thermal stability, high temperature strength, good ablation characteristics, and good insulating capability. Since graphite possesses greater heat resistance than carbon and is physically stable at elevated temperatures, graphite reinforcements are used in rocket nozzle throats and ablation chambers. Carbon-carbon composites are used when greater strength and lower conductivity is required.

Tailoring. Composites often are considered as lightweight alternatives for more traditional structural materials. Innovation with composites can lead to higher production efficiency and lower life-cycle costs. For example, creativity in using raw materials allows the product design-to-production team to build the composite at the same time the structure is fabricated and to use the anisotropy of the materials to tailor the properties of the composite to meet specific structural requirements.

Fabrication. Current composites fabrication techniques are similar to those used for producing fiberglass, including hand and automated tape lay-up resin injection, compression molding, vacuum-bag and autoclave molding, matched die molding, pultrusion, and filament winding. Organic matrix composites are made primarily by molding in autoclaves, while metal (aluminum, titanium, etc.) matrix composites are formed mainly by diffusion bonding.

Lamination, filament winding, pultrusion, and resin-transfer (injection) molding are four widely used methods of producing continuous-fiber composites with closely controlled properties. The shape, size, and type of the part and the quantity to be manufactured determine construction techniques. The lamination method is used for comparatively flat pieces. Filament winding is a powerful and potentially high-speed process for making tubes and other cylindrical structures. So-called pultrusion and pulmolding can be used for parts with constant cross-sectional shapes. Injection molding can be used for small, nonload-bearing parts. An epoxy injection process, called URTRI (ultimately reinforced thermoset reaction injection), is used for making load-bearing structures, sandwiches, and torsion boxes.

Laminating. Advanced composites are typically used in the form of laminates and are processed by starting with a prepreg material (partially cured composite with the fibers aligned parallel to each other). A pattern of the product's shape is cut out, and the prepreg material is then stacked in layers into the desired laminate geometry.

A final product is obtained by curing the stacked plies under pressure and heat in an autoclave. Graphite-epoxy composites are cured at approximately 350°F (175°C) at a pressure of 100 psi (690 kPa). The new high-temperature composites, such as bismaleimides, are cured at 600°F (316°C). The tooling is essentially a mold that follows a part through the lay-up and autoclaving processes. Tooling materials commonly used for manufacturing composite parts include aluminum, steel, electroplated nickel, a high-temperature epoxy-resin system casting, and fabricated graphite composite tools.

Filament Winding. In the filament-winding process, fibers or tape are drawn through a resin bath and wound onto a rotating mandrel. Filament winding is a relatively slow process, but the fiber direction can be controlled and the diameter can be varied along the length of the piece. In some versions of the process, the fiber bundle, which may be made up of several thousand carbon fibers, is first coated with the matrix material to make a prepreg tape. The tape is an endless strip with a width that may vary from an inch to a yard (several centimeters to a meter). With both the fiber and tape-winding processes, the finished part is cured in an autoclave and later removed from the mandrel.

For strength-critical aerospace structures, carbon fibers are usually wound with epoxy-based resin systems. The polyesters, phenolics, and bismaleimides are limited to special applications. Filament winding is used to produce round or cylindrical objects such as pressure bottles, missile canisters, and industrial storage tanks. It also has been used to make automobile drive-shafts.

Pultrusion. In composites technology, pultrusion is the equivalent of metals extrusion. Pultrusion (also called pultruding), consists of transporting a continuous fiber bundle through a resin matrix bath and then pulling it through a heated die. The process can be used to make complex shapes; however, it has been limited to items with constant cross sections, such as tubing, channels, I-beams, Z-sections, and flat bars. Developmental activity is progressing on variable-section pultrusion, in which the geometry can be controlled by an articulating die. Pulmolding is a process variation that begins with pultruding; then the part is placed in a compression mold.

Resin Transfer Molding. Filling a niche between hand manufacturing lay-up or spray-up of parts and compression molding in matched metal molds is resin transfer molding (RTM), also called resin injection molding.

In the conventional RTM process, two-piece matched cavity molds are used with one or multiple injection points and breather holes. The key to RTM is low pressure in the mold, which allows the use of low-cost tooling. The reinforcing material, either chopped or continuous strand mat, is cut to shape and draped in the mold cavity. The mold halves are clamped together, and a polyester resin is pumped through an injection port in the mold. Polyester, glassmat, and conventional RTM are not considered to be in the family of composites.

Compared with the spray-up method, RTM permits faster cycle times and usually requires less labor. The RTM cycle times are longer than for compression molding, but the low cost of RTM tooling often compensates for the differential when the production run is fewer than 50,000 parts per year.

In the advanced RTM process, URTRI, reinforcements are placed in the mold and cores can be handled as inserts. In this process, a core is cast from synthetic foam (high-temperature epoxy with hollow glass microspheres) around a fitting. The core is then wrapped with multiple layers of bidirectional graphite fabric in 7- and 14-mil (0.18- and 0.36-mm) thicknesses and placed in the mold. Epoxy resin is next injected into the heated mold. Curing time is five minutes.

Potential benefits of URTRI include reduced part weight and cost and improved quality for items such as aircraft wings, fins, elevons, passenger seat shells, landing-gear beams, and other high-performance structures.

Practice Problems

17.1 What type of information does the following chart provide the design engineer?

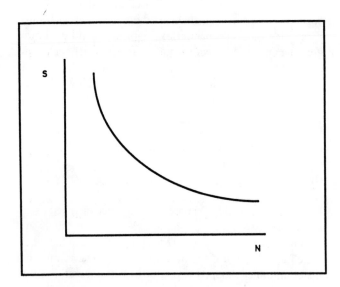

17.2 What properties of a material are revealed in a tensile test?

17.3 In the design process, the engineer must take into account three important properties of metal. What are they?

17.4 Give a brief description of the three properties of question 17.3.

17.5 When steel is surface (cure) hardened, what new characteristic will it exhibit?

17.6 Which of the following treatments of metal is not used for surface hardening?

    a. Carburizing
    b. Cyaniding
    c. Nitriding
    d. Normalizing

17.7 Which of the following is not a property of plastics as defined by the American Society of Testing and Materials?

    a. Organic
    b. Solid in its finished state
    c. Can be shaped by flow
    d. Has a small molecular chain

17.8 Advanced composites are distinguished from reinforced plastics in that

    a. They use fibers in a resin binder for strength
    b. Their polymer resin matrices are thermoset only
    c. Their fiber-to-resin ratio by weight is greater than 50%
    d. They can be machined dry

STRATEGIC ISSUES

18.1 Manufacturing Strategy

To be successful in product development, different organizations have adopted various strategies to accommodate fast-changing markets. Many firms in the 1990s believe that initiation and innovation strategies are most likely to succeed rather than imitation. Some of these strategies include, but are not limited to:

- Customer responsive strategy,

- Entrepreneurial manufacturing strategy,

- Time-based strategy, and

- Managing for speed product strategy.

Common to all is the objective of producing a product at a competitive price. Firms with a traditional mass-production manufacturing strategy acknowledge that to remain competitive, they must develop a flexible specialization strategy that enables them to be more entrepreneurial in their manufacturing approach by both accommodating and creating rapid market changes.

Customer Responsive Strategy. A customer responsive strategy targets quality improvement and customer service. It integrates an effective organizational structure with good human resource management and efficient production processes. The manufacturing function plays a big role in this approach and cultivates employee involvement and departmental partnering, where internal cooperation, not competition, is what is necessary to succeed. This scheme sets the stage for short-run manufacturing to be attained through the utilization of the work cell concept. By gaining this flexibility, and by implementing total quality control, companies have combined the necessary elements to produce products in just-in-time (JIT) mode and capture a competitive edge.

Entrepreneurial Manufacturing Strategy. The entrepreneurial manufacturing strategy is based on a concept similar to that of customer responsiveness. It requires flexible manufacturing capable of shifting from one product to another on short notice. It provides a continuous stream of new products to specialized markets and creates an integrative organizational structure to allow smooth operation across functional activities. However, it does instill in members an entrepreneurial spirit built on pride, commitment, collaboration, and teamwork. The success of this strategy is based on a company's capacity to create new markets for specialized high-value-added products rather than continuing with standardized products at lower prices.

Time-based Strategy. The dynamics of a time-based strategy are designed to achieve high productivity and low cost. Fundamental to its success is its focus on offering variety rather than volume. This approach is founded on three basic elements:

1. Organization of process components and standardization,

2. Length of production run, and

3. Complexity of scheduling procedures.

This strategy favors smaller increments of improvement in new products but introduces them more often. This product development work uses factory cells staffed by cross-functional teams and stresses local responsibility in scheduling. Figure 18.1 provides a comparison between traditional and flexible manufacturing.

Managing for Speed Product Strategy. All aspects of the strategy that manages for speed to market revolve around targeting speed as first and foremost, including:

1. Organizing product development for speed,

2. Organizing product manufacturing for speed,

3. Using miscellaneous techniques for speed, and

4. Using computer-aided technology for speed.

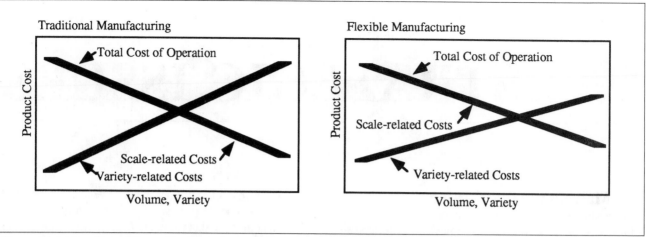

Figure 18.1 Cost Comparison, Traditional versus Flexible Manufacturing.

This strategy requires that all departments be in proximity to one another, and depends on individual discipline as well as team effort to ensure simultaneous consideration of all interfunctional requirements. All possible modern manufacturing techniques and computer aids are taken into account to enhance process speed. In addition, broader task orientation and up-to-date skills are required to accommodate the flexibility of the system.

All strategies described focus on delivering a quality product at a competitive price, simultaneously responding to customer needs, and striving for continuous improvement.

However, these strategies are not packaged, off-the-shelf items offering turnkey implementation. With each, a feasibility study to ensure applicability is needed. Implementation without planning could drive up costs and leave a negative impact that may be detrimental to the existence and survival of such companies.

18.2 **Common Techniques**

Since the early 1980s, foreign competition has escalated pressure on American companies on their own turf by offering various products within shorter cycles. However, American firms, with their sequential product development under scrutiny, quickly realized that change was in order. As a result, techniques such as:

• Concurrent engineering,

• Computer-aided applications, and

• Rapid prototyping

became more common among adapted strategy. These techniques helped companies reduce their cost and achieve shorter cycles to face the competition.

Concurrent Engineering. By definition, concurrent engineering is designing a product and designing the process to manufacture that product during the concept phase. It uses many disciplines that cooperatively achieve an acceptable and manufacturable design on the first iteration. By so doing engineering changes are minimized, product cost is reduced, and the overall development cycle is shortened. Concurrent engineering has a lasting effect on the life cycle of the product. It brings all engineering operations together and creates an environment in which efforts are combined as a single entity to attack problems early in the design stage when changes are least costly. This approach encourages open communication and creates camaraderie among team members. Honeywell, Boeing, and the Big Three automotive companies are representative of many companies that have successfully implemented concurrent engineering. Some of the tools of the concept include quality function deployment (QFD), design for manufacturability (DFM), design for assembly (DFA), and advanced design and evaluation analyses.

Quality Function Deployment (QFD). QFD focuses on the customer to ensure that everyone is well aware of exactly what customers need and why they need it. In addition, ergonomics are considered and designs are fitted to suit human factors from a fifth percentile female to a 95th percentile male, therefore covering all ranges. Traditionally, product design was a chain reaction that started with Sales/Marketing, moved to Product Planning, Design Engineering, Manufacturing, and finally to Distribution. However, the resulting interpretation rarely bears familiarity with the original needs.

On the other hand, QFD seats all cognizant parties at the table on a regular basis to simultaneously discuss customer requirements. Members of such a team usually represent:

- Manufacturing,

- Engineering,

- Sales/Marketing,

- R&D,

- Purchasing, and

- Quality Assurance.

This forum allows parallel development of all involved activities and clearly defines customer needs and wants. The benefits are numerous, not the least of which are greater cooperation, higher productivity, and increased profitability.

Design for Manufacture/Design for Assembly (DFM/DFA). Several versions of DFM/DFA are available to manufacturers today. At the core of each, however, are four activities, each addressing particular aspects of the DFM approach.

1. Optimize Product/Process Concept.

2. Simplify Product Design.

3. Ensure Product/Process Performance.

4. Optimize Product Function.

This arrangement, as shown in Figure 18.2, emphasizes the iterative nature of the process. Some of the tools of DFM are group technology, value analysis, failure mode and effect analysis (FMEA), Taguchi methods, computer-aided technology, Boothroyd-Dewhurst DFA method, etc.

Attention, communication, and cooperation are key to the success of the DFM approach.

Advanced Design and Evaluation Analyses. The evolution of technology in the late twentieth century helped manufacturing take a giant leap toward modernization and automation. Development of computer-aided applications for a variety of manufacturing functions provided companies with the digital power and speed to compete efficiently in global markets. Such applications today pervade the industrial enterprise, beginning with product design and modeling and carrying through to aftermarket support.

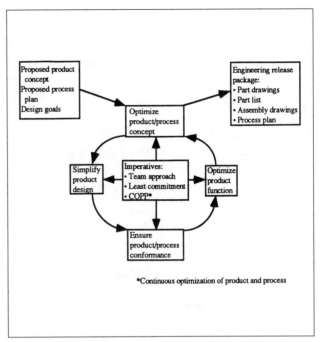

Figure 18.2 Activities of a Typical DFM/DFA Process.

These modeling, simulation, and risk analysis techniques have ushered in a bold new era of product development. They not only reduce the number of engineering changes but push them upstream where they can be made much more easily and at far less expense. Some practical examples to better acquaint the reader with such analyses include:

- Fault-tree Analysis is a quantitative method using deductive reasoning and Boolean algebra to identify critical paths of an undesired event.

- Finite Element Analysis is used to evaluate such events as crush sequence of components in a barrier test or an air bag deployment or to model heat distribution in an engine exhaust manifold.

- On the process side, various computer simulations use meshing/mapping techniques to evaluate the manufacturability of components by analyzing stress/strain potential failure areas, etc.

Computer-aided Applications. The need for shorter product development cycles has driven companies to focus on computer systems to automate design and processes. As a result, many software packages were developed independently to suit specific needs and objectives. However, because of the proprietary nature of the formats of such software, system-to-system compatibility became a real problem when electronic transfer of information was needed. Therefore, a joint government/industry force was created in 1979 to develop a method for data exchange. Through their efforts, the Initial Graphics Exchange Specification (IGES) was published in 1980 to facilitate data exchange independent of CAD/CAM systems. The IGES methodology enables manufacturers to exchange CAD data between different CAD systems. Figure 18.3 shows how the process works.

CADD, CAD/CAM, CAE, CIM, and a host of other computer-driven technologies have provided industry with tremendous time savings and flexibility. Development of these systems is one of the most significant engineering breakthroughs in decades.

Yet, for all the advantages and benefits of computer-aided systems, the importance of integrating design and manufacturing engineering functions cannot be over-emphasized. At a time when time-to-market emerges as critical to product success, concurrent engineering emerges as critical to compressing the time-to-market cycle. In the future, mailing of blueprints will become obsolete as data flows electronically from system to system and part definition is secured in digital form on computer tape or disk.

Rapid Prototyping. Rapid prototyping uses modern technology to produce sample parts in a very short time for engineering evaluation. The evolution of technology made it possible to create different methods/techniques such as:

- Stereolithography (STL),

- Solid ground curing (SGC),

- NC/CNC machining,

- Investment casting,

- Laminated object manufacturing (LOM),

- Fused deposition modeling (FDM),

- Selective laser sintering (SLS),

- Ballistic particle manufacturing (BPM),

- Epoxy mold, and many more.

Figure 18.4 lists some rapid prototyping applications.

Rapid Prototyping Capabilities, Concepts, and Tools. Two strong capabilities of rapid prototyping are:

1. It can automate the creation of working designs and plans,

2. It can generate 3-D graphics to make visualization easier and provide easy execution of various engineering analyses.

It has three basic concepts:

1. Provides tools, generates and displays shapes;

2. Relieves engineers of dimensioning and scaling;

3. Provides instant availability for further manipulation.

Some rapid prototyping features are:

- Capability of rotating the view of the drawing and creating partial blowup of a drawing,

- GDT representation to facilitate gaging and QC functions, and

- Circular array to create evenly spaced holes.

Stereolithography. STL uses a computer file to produce a desired part on a computer screen. This technique accepts a 3-D solid model from a CAD system. The process of producing the part is done by a sequential build using a laser of successive thin-sliced layers of epoxy until the part takes its final shape. This commonly produces complex plastic parts directly from a 3-D CAD model.

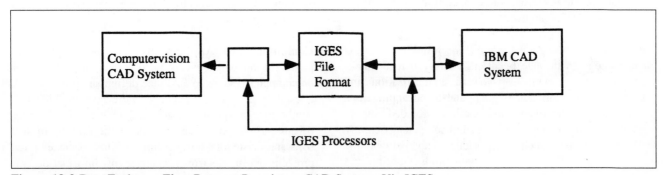

Figure 18.3 Data Exchange Flow Between Proprietary CAD Systems Via IGES.

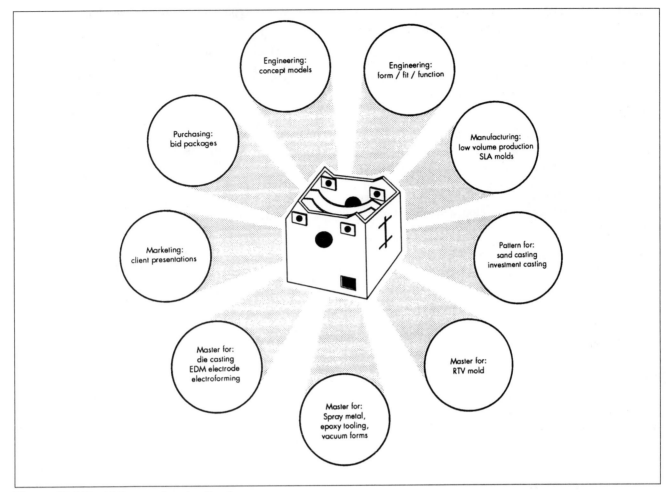

Figure 18.4 Rapid Prototyping Applications.

Selective Laser Sintering. SLS is similar to STL except the part is not created in a liquid vat but from a heat-fusible wax powder that is spread over the part piston.

Advantages and Disadvantages of Rapid Prototyping. Among the advantages of rapid prototyping are:

- Speed in part realization,

- Capability of actually building the part's mold,

- Elimination of cost- and time-intensive early prototypes, and

- Debugging of early design flaws.

However, it has some disadvantages and limitations as well:

- Limits development of full-scale models,

- Parts typically cannot be used for physical testing,

- Special techniques and materials are required of some systems,

- Limits manufacturing experience with prototypes,

- Is a new technology,

- Has surface finish and tolerance limitations, and

- Has inherent limitation in wall thicknesses and constraints on feature configuration, etc.

161

Practice Problems

18.1. What are the three basic elements of time-based strategy?

18.2. At what stage in product development is it least costly to make product changes?

    a. Design
    b. Manufacturing
    c. Sales
    d. Customer Service

18.3. Which specification was created jointly by government and industry to develop a method for data exchange?

    a. IGES
    b. DFM
    c. QFD
    d. MAP/TOP

18.4. Which is not a rapid prototyping method:

    a. Investment casting
    b. Laser sintering
    c. Forging
    d. Stereolithography

18.5. A rapid prototyping method that uses a laser and an epoxy to create a plastic-like model part is called:

    a. Laser beam welding
    b. Clay modelling
    c. Stereolithography
    d. NC machining

MANUFACTURING PROCESSES

19.1 Processes and Process Planning

Process planning and process design cover many facets of the manufacturing enterprise, facets independent of the end product. Regardless of whether the products are mechanical, electrical, or chemical, planning and design steps are essentially the same. Some factors affecting process design are:

- The volume of the product to be manufactured,

- The quality specifications,

- The equipment available or attainable for manufacturing.

Methods engineering and work measurement form another component of process planning. Methods engineering focuses on analyzing methods and equipment used in performing a task, on either existing processes or future jobs.

Several tools are available for methods engineering such as process charts, micromotion, and memomotion. Process charts are a graphic representation of the step-by-step sequence taking place in the manufacturing cycle. With the emergence of motion pictures and video recorders, filming workers performing their tasks became a popular method of analyzing job performance. Micromotion uses motion pictures taken at constant and known speeds. It is applicable for analyzing processes with short cycle times and rapid movements. Memomotion, on the other hand, uses a slower film speed and is quite useful in analyzing jobs with long cycle times or those involving many interrelationships.

Work measurement is another process design tool. By standardizing or allowing times for specific tasks, engineers plan and schedule production and conduct cost estimating and line balancing. During the process of making a time study, the operator may work faster or slower than what the analyst considers to be normal. To compensate, it is necessary for the analyst to rate the operator. Essentially, this means that the analyst compares the performance of the operator with the analyst's opinion of normal performance. The rating factor is expressed as a percentage of normal performance (normal performance = 100%). Normal time can be calculated by the following:

Normal time = average time × percent rating/100

With machine down time, material delays, and other interruptions, the normal time is not completely accurate. To compensate for delays and interruptions, a standard time is calculated by increasing the normal time by an allowance. The standard time is calculated as:

Standard time = normal time × (100 + allowance in percent)/100

Example 19.1.1. After 15 trials, the average time for a worker to assemble a cable clamp is 10.3 seconds. If the rating factor is 95, what is the normal time?

Solution. Normal time = 10.3 seconds × 95/100
= 9.8 seconds

Example 19.1.2. Due to assembly fatigue and humid conditions, an allowance of 10% is necessary. Calculate the standard time for the process in example 19.1.1.

Solution. Standard time = 9.8 seconds × (100 + 10)/100
= 10.78 seconds

19.2 Process Equipment Controls

There are several types of process equipment, including fixed power tools and equipment, portable power tools, hand tools, and auxiliary tools. In general, portable power tools, hand tools, and auxiliary tools are controlled directly through human coordination; that is, they are not programmed. Fixed power tools and equipment, however, are controlled through some type of control program. These controls consist of the following:

- Sequence controls,

- Programmable logic controllers (PLC),

- Numerical control,

- Adaptive control.

Sequence controllers are a class of electromechanical and electronic devices to control a machine tool operation or other equipment. The more common types of sequence controllers are electromechanical stepping-drum programmers and perforated paper-tape programmers.

In the drum programmer, the desired control sequence is commonly established by inserting pins into appropriate rows in the cylinder surface. Mounted over one row of the cylinder surface are momentary contact switches. As the cylinder rotates, the pins in a row activate input devices, such as pushbuttons or timer contacts. The logic section causes the closure of circuits to the output devices. When the logic section senses that all the outputs are in the correct condition for that row, it advances, or ''steps,'' to the next row.

Perforated paper-tape programmers rely on a pattern of punched holes in a paper tape similar to an old player piano.

A PLC is a solid-state device used to control machine motion or process operation with a stored program. The PLC sends output control signals and receives input signals through input/output (I/O) devices. Input is made through limit switches, pushbuttons, etc. Output devices consist of solenoids, motor starters, etc. The processor part of a PLC is the central processing unit (CPU). The CPU scans the state of I/O and updates outputs based on instructions stored in the memory of the PLC.

Numerical control or computer numerical control machine tools and other manufacturing systems use prerecorded, written symbolic instructions. Numerical control machines use G-codes and M-codes, automated programmed tool (APT), or other instructions to control the machine tool. NC control systems can be either open-loop or closed-loop with the latter being the preferred system. Open-loop means that the machine activators receive and execute signals from the controller but there is no feedback to the controller. In

a closed-loop system, the activators provide feedback to the controller after the signals have been executed.

Adaptive control (AC), sometimes referred to as automatic adaptive control (ACC), automatically and continuously identifies on-line performance of an activity (a process or operation, for example). AC measures one or more variables of the activity, compares the measured quantities with other measured quantities, calculates quantities or established values or limits, and modifies the activity by automatically adjusting one or more variables to improve or optimize performance.

Maintenance is an important issue in manufacturing. There are three basic types of maintenance:

- Corrective,

- Preventive,

- Predictive.

Corrective maintenance is simply fixing equipment when it breaks down. Preventive maintenance concerns itself with performing maintenance activities on equipment to prevent larger maintenance problems from occurring later. Finally, predictive maintenance concerns itself with predicting when equipment will need maintenance. By using sensitive instruments such as vibration analyzers, imminent equipment failure can be predicted and avoided.

19.3 Assembly

Five basic types of assembly systems are in use today: single-station assembly, synchronous assembly, nonsynchronous assembly, continuous-motion assembly, and dial (rotary) assembly.

Single-Station Assembly. Machines having a single workstation are used most extensively when a specific operation must be performed many times on one or a few parts. These machines may also be used when different operations must be performed and if the required tooling is not too complicated. These machines are incorporated into multistation assembly systems.

Synchronous Assembly. Synchronous (indexing) assembly systems are available in dial, in-line, and carousel varieties. With these systems, all pallets or workpieces are moved at the same time and/or the same distance. Synchronous systems are used primarily for high-speed and high-volume applications on small lightweight assemblies where the various operations have relatively equal cycle times.

Nonsynchronous Assembly. Nonsynchronous transfer (accumulative or power-and-free) assembly systems, with free or floating pallets or workpieces and independently operated individual stations, are being widely used where the times required to perform different operations vary greatly. The systems are also applied for larger products having many components. One major advantage of these so-called power-and-free systems is increased versatility.

Continuous-motion Assembly. With continuous-motion systems, assembly operations are performed while the workpieces or pallets move at a constant speed and the workheads reciprocate. High production rates are possible because indexing time is eliminated. However, the cost and complexity of these systems are high as workheads have to synchronize and move with the product being assembled.

Dial (Rotary) Assembly. Dial or rotary index machines of synchronous design are one of the first types used for assembly. They are still used for many applications. Workstations and tooling can be mounted on a central column or around the periphery of the indexing table. These machines are generally limited to small and medium-sized lightweight assemblies requiring a relatively low number of operations that are not too complicated.

19.4 Mechanics of Chip Formation

Tool Nomenclature. Solid single-point cutting tools can be made of high-speed steels or a variety of carbides. Brazed tip single-point cutting tools means the cutting material is brazed onto a less expensive material for the tool body. The nomenclature is the same for both solid and tipped tools. A single-point tool (Figure 19.1) embodies several geometrical elements:

- Size—The size of a tool of square or rectangular section is expressed by giving, in the order named, width of shank, W; the height of shank, J; and total tool length, L, in inches (mm), such as 3/4 × 1½ × 8 (19 × 38 × 203 mm).

- Shank—This is the part of the tool by which it is held.

- Base—This is a flat surface on the tool shank, parallel or perpendicular to the tool reference plane and useful for locating or orienting the tool in its manufacture, sharpening, and measurement.

- Heel—This consists of the area adjacent to the intersection of the base and flank.

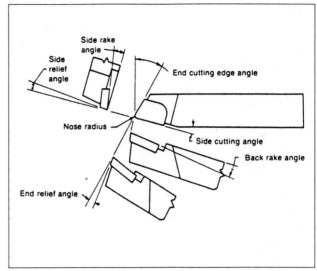

Figure 19.1 A Single-point Cutting Tool.

- Face—This is the surface the chip contacts as it is separated from the workpiece.

- Tool point (cutting part)—This is the part of the tool shaped to produce the cutting edges, face, and flank.

- Cutting edge—This is the portion of the face edge along which the chip is separated from the workpiece. It usually consists of the side cutting edge, the nose, and the end cutting edge.

- Nose—This is the corner, arc, or chamfer joining the side cutting and the end cutting edges.

- Flank—This is the surface or surface below and adjacent to the cutting edge.

- Neck—This is an extension of the shank that has a reduced sectional area. A relatively small point, as required in boring, is sometimes attached to the shank by a neck.

Tool Angles. The angles shown in Figure 19.1 are "normal," i.e., taken with reference to the cutting edges, as these are the ones specified in grinding a single-point tool. Tool angles include:

- Back rake angle—This is the angle between the face of the tool and the base of the shank or holder, usually measured in a plane through the side cutting edge and at right angles to the base. It is positive if the face slopes downward from the point toward the shank, tending to reduce the included angle of the tool point. It is negative if the face slopes upward toward the shank.

- Side rake angle—This is the angle between the face of the tool and a plane through the cutting edge, parallel to the base of the shank or holder. It is usually measured in a plane to the base and perpendicular to the side cutting edge; hence, it is a normal side rake angle.

- Side relief angle—This is the angle between the portion of the side flank immediately below the side cutting edge and a line drawn through this cutting edge perpendicular to the base. It is usually measured in a plane at right angles to the side flank; hence, it is a normal side relief angle.

- End relief angle—This is the angle between the portion of the end flank immediately below the end cutting edge and a line drawn through that cutting edge perpendicular to the base. It is usually measured in a plane at right angles to the end flank; hence, it is a normal end relief angle.

- Clearance angle—This is the angle between a plane perpendicular to the base of the tool or holder and that portion of the flank immediately below the relieved flank. Side clearance angle is measured in the plane of the side rake angle. The clearance angle is greater than its corresponding relief angle, except when only one plane exits on the flank, in which case, the clearance and relief angles would coincide.

- Side cutting edge angle—This is the angle between the side cutting edge and the side of the tool shank or holder.

- End cutting edge angle—This is the angle between the cutting edge of the end of the tool and a line at right angles to the side of the tool shank.

Working Angles. Working angles are the angles between the tool and the work and depend not only on the shape of the tool but also on its position with respect to the work. Working angles include:

- Setting angle—This is the angle made by the straight portion of the shank of a tool or holder with the machined portion of the work (most commonly 90°).

- Entering angle—This is the angle that the side cutting edge of a tool makes with the machined surface of the work (90° in the case of a tool with an effective 0° side cutting angle).

- True rake angle—This is the slope of the tool face toward the tool base, from the active cutting edge in the direction of chip flow.

- Cutting angle—This is the angle between the face of the tool and a tangent to the machined surface at the point of action. It equals 90° minus the true rake angle.

- Lip angle—This is the included angle of the tool material between the face and the relieved flank. According to directions of measurement, it may represent the end lip, side lip angle, or the true lip angle.

- Work relief angle—This is the angle between the relieved ground flank of the tool and a line tangent to the machined surface passing through the active cutting edge.

- Working-end cutting-edge angle—This is the angle between the end cutting edge and a plane tangent to the machined surface at the point of cutting.

Orthogonal Cutting Model. The thickness of the chip, t_c, is always greater than the undeformed chip thickness, t, in orthogonal cutting. The ratio of t to t_c is called the cutting ratio, r (Figure 19.2).

In orthogonal cutting, the width of the chip, b_c, equals the width of the work, b, to a good approximation (as long as $b/t \leq 5$). When any metal is deformed plastically, no change occurs in volume; hence;

$$lbt = l_c b_c t_c$$

or since the width of the work, b, equals the width of the chip, b_c,

$$r = t/t_c = l_c/l$$

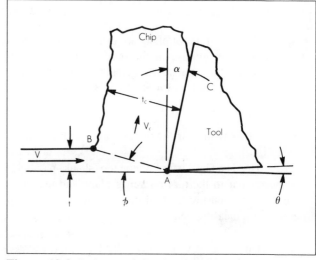

Figure 19.2 Orthogonal Cutting Model.

where:

l = undeformed chip length (in. or mm)

l_c = corresponding chip length (in. or mm).

The cutting ratio is a convenient measure of cutting efficiency. Figure 19.3 shows material being orthogonally cut with high and low cutting ratios and corresponding shear angles ϕ_1 and ϕ_2. Since most of the energy in orthogonal cutting is associated with the shear plane, less energy is required to form thin chips than thick ones.

One of the important uses of the cutting ratio, r, is in estimating the shear angle, ϕ. This may be done by the use of the following equation:

$$\tan\phi = r\cos\alpha/1 - r\sin\alpha$$

Example 19.4.1. Determine the shear angle, ϕ, if the cutting ratio is 0.51 and the rake angle is 10 degrees.

Solution. ϕ = arctan $0.51\cos10/(1-0.51\sin10)$
$\qquad$ = 26.86 degrees

Oblique Cutting Model. When the cutting edge deviates from the orthogonal direction, the angle of deviations is called the inclination angle, i. Figure 19.4 shows a tool with an inclination angle, i. The effect of this angle is to cause the chip to flow up the tool face in a direction different from that of orthogonal cutting. In orthogonal cutting, the chip flows up the tool face in the direction of a normal angle to

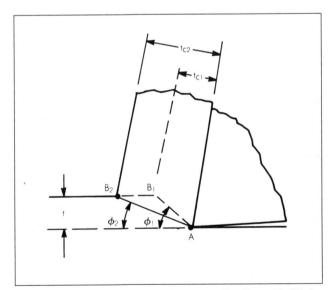

Figure 19.3 Material Being Orthogonally Cut with High and Low Cutting Ratios.

the cutting edge, n. However, when the inclination angle, i, is not equal to 0°, the chip flows at an angle n_c to the normal to the cutting edge measure in the plane of the tool face (Figure 19.4). Thus, V_c is at an angle n_c to normal, n, when i is not equal to 0. To a reasonably good approximation, the inclination angle, i, measured in the plane of the uncut surface, is equal to the chip flow angle, n_c, measured in the plane of the tool face. This is known as Stabler's Rule and is expressed as:

$$n_i = i$$

The effective rake angle α, is measured in the plane of V and V_c and differs from the normal rake angle, α_n, measured in the plane containing V and the angle normal to the cutting edge, n. For the general case when i is not equal to 0, the effective rake angle, α, is less than the normal rake angle, α_n.

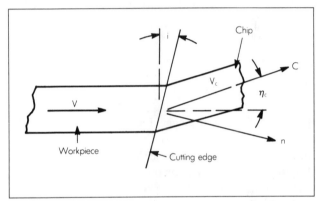

Figure 19.4 Plan View of Tool with Nonzero Inclination Angle.

19.5 Turning

Lathes. Lathes and turning machines come in a wide variety of types in many sizes to suit specific application requirements. Engine lathes, contouring lathes, turret lathes, and NC/CNC turning machines are just a few of the types in use today.

The parts of an engine lathe can be seen in Figure 19.5.

Engine lathes are generally used for tool production and low-volume manufacturing runs. They are capable of facing, straight turning, parting, boring, and thread cutting. Generally, engine lathes use single-point tools, parting tools, thread-cutting tools, boring bars, and form tools. Work holding devices consist of three-jaw universal chucks, four-jaw independent chucks, and collets. Typically, three-jaw universal chucks are used when tolerances are not high and the workpiece geometry will permit it. Four-jaw inde-

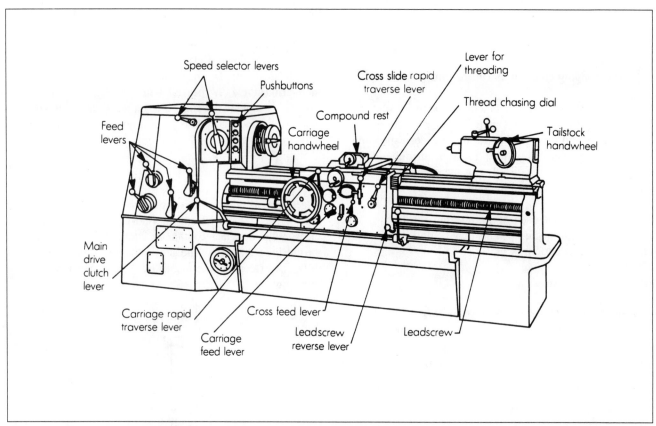

Figure 19.5 Parts of an Engine Lathe.

pendent chucks are used when the workpiece geometry does not allow a three-jaw. Collets provide maximum accuracy but are limited to round workpieces.

Turning Calculations. The revolutions per minute (rpm) for turning operations can be calculated from the following formulas.

Metric:
rpm = (cutting speed × 1000) / (D × π)

Inch:
rpm = (cutting speed × 12)/(D″ × π)

Example 19.5.1. What rpm should be used for turning a piece of low-carbon steel of 25.4-mm diameter with a cutting speed of 30.5 mpm?

Solution. rpm = (30.5 × 1000)/(25.4 × π)
= 382 rpm

Example 19.5.2. Calculate the motor horsepower requirements for machining 2-in.-round 1020 steel with a depth of cut of 0.060 in., feed rate of 0.00072 in./rev, 75% efficiency, 100 ft/min cutting speed, and unit horsepower of 1.

Solution. Q = 12 × 2 in. ×0.00072 in./rev × 100 ft/min
= 1.728 in.^{3}/min

HP_s = 1.728 in.^{3}/min × 1 hp/in.^{3}/min
= 1.728 hp

HP_m = 1.728hp/0.75
= 2.34 hp

19.6 Drilling

Drill Machines and Drills. Drilling is the production of holes by the relative motion of a rotating cutting tool and the workpiece. Drilling machine types include vertical, gang, radial, turret, and screw (small holes).

Twist drills are a common type of cutting tool used for drilling. Figure 19.6 illustrates the parts of a twist drill.

Drilling Calculations. To calculate cutting speed, follow the formula below:

Metric:
rpm = CS (mpm) × 1000/ D(mm) × π

168

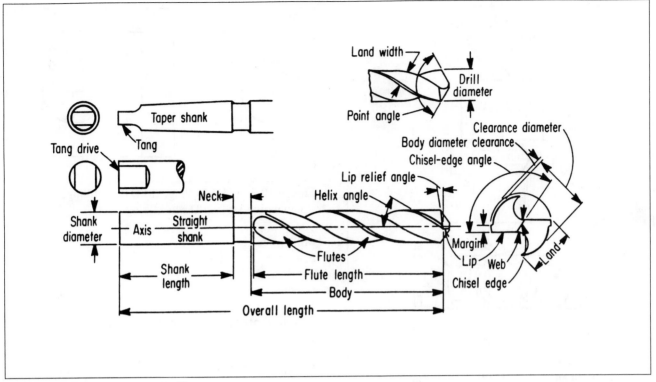

Figure 19.6 Parts of a Twist Drill.

Inch:

$$rpm = CS \text{ (fpm)} \times 12/ D\text{(in)} \times \pi$$

Example 19.6.1. Calculate the rpm for a 0.5-in. drill cutting through 1-in. aluminum with a cutting speed of 300 fpm.

Solution. rpm $= (300 \text{ fpm} \times 12)/(0.5 \times \pi)$
$\qquad\qquad = 2293$ rpm

Some forces and power in drilling are related in Figure 19.7.

Example 19.6.2. Calculate the motor horsepower requirements for drilling a 1-in. hole in low-carbon steel at 100 fpm cutting speed, 0.00052 in./rev feed rate, 75% efficiency, and unit horsepower of 1 at 700 rpm.

Solution. Q $= \pi/4 \times 1$ in. $\times 0.00052$ in./rev $\times 100$ fpm
$\qquad\qquad \times 700$ rpm
$\qquad\quad = 0.286$ in.$^{3-}$

$\qquad$ HP$_s$ $= 0.286$ in.$^{3-}$ $\times 1$
$\qquad\qquad = 0.286$ hp

$\qquad$ HPm $= 0.286/0.75$
$\qquad\qquad = 0.381$ hp

19.7 Milling

Milling Machines. Milling is a machining process for removing material by relative motion between a workpiece and a rotating cutter having multiple cutting edges. In some applications, the workpiece is held stationary while the rotating cutter is moved past it. In other situations, the rotating cutter is held stationary while the workpiece is moved into it.

The types of milling machines consist of standard vertical or horizontal knee-and-column machines, CNC milling machines, and machining centers.

General milling methods consist of end milling, straddle milling, gang milling, cam milling, and thread milling. There are two possible ways of performing these milling methods, up milling and down milling (Figure 19.8). In up milling, the tangential cutting force opposes the direction of the workpiece feed. In down, or climb, milling, the tangential cutting force is in the direction of the workpiece feed.

Up and down milling exist in their pure form only when the cutter spindle centerline does not intersect the workpiece. In up milling, the feed force is high because the tangential force is attempting to push an individual tooth or insert out of the cut. In down milling, however, tangential forces act in the same direction as the feed, therefore decreasing the feed force.

169

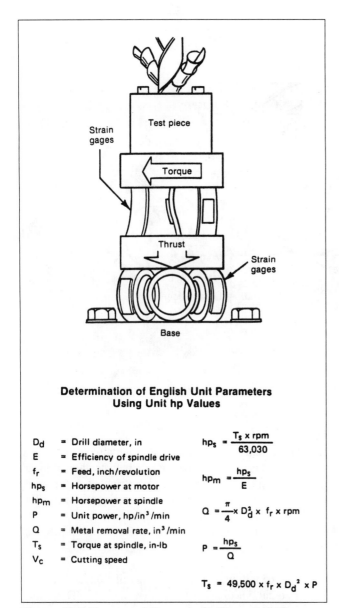

Determination of English Unit Parameters Using Unit hp Values

D_d = Drill diameter, in

E = Efficiency of spindle drive

f_r = Feed, inch/revolution

hp_s = Horsepower at motor

hp_m = Horsepower at spindle

P = Unit power, hp/in³/min

Q = Metal removal rate, in³/min

T_s = Torque at spindle, in-lb

V_c = Cutting speed

$$hp_s = \frac{T_s \times rpm}{63,030}$$

$$hp_m = \frac{hp_s}{E}$$

$$Q = \frac{\pi}{4} \times D_d^2 \times f_r \times rpm$$

$$P = \frac{hp_s}{Q}$$

$$T_s = 49,500 \times f_r \times D_d^2 \times P$$

Figure 19.7 Forces and Power in Drilling. (*IAMS*)

Down or climb milling is preferred when the setup permits. With down milling, chip thickness at entry is at a maximum and any work-hardened layer is avoided. Also, when down milling is used any welded chips are generally cut in half as the tooth or insert enters the cut (Figure 19.8).

Milling Calculations. Milling cutting speed is determined as follows:

Metric:
cutting speed = D(mm) $\times \pi \times$ rpm/ 1000

Inch:
cutting speed = D(in.) $\times \pi \times$ rpm/ 12

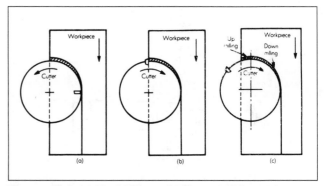

Figure 19.8 (a) Up Milling, (b) Down Milling, (c) Combination Up and Down Milling.

Milling feed rate
$F = R \times T \times$ rpm

where:
F = feed rate, mm/min. or in./min.
R = feed per tooth per revolution, mm or in.
T = number of teeth

Example 19.7.1. Determine the cutting speed of an 8-inch cutter turning at 100 rpm.

Solution. Cutting speed = 8 in. $\times \pi \times$ 100 rpm/12
= 209 fpm

Some forces and power in milling are shown in Figure 19.9.

Example 19.7.2. Determine the spindle horsepower needed for a horizontal mill, if a 3-inch by 1-inch cutter with 8 teeth is making a 0.250-in. deep cut at 120 rpm, with a feed rate of 0.050 in./tooth and unit horsepower of 1.

Solution. Q = 1 in. $\times$ 0.250 in. $\times$ 0.50 in./tooth $\times$ 8 teeth
$\times$ 120rpm
= 12 in.³/m

HP_s = 12 in.³/m $\times$ 1
= 12 hp

19.8 Bandsawing

Power bandsawing uses a long endless band with many small teeth traveling over two or more wheels (one is a driven wheel and the others are idlers) in one direction. The cutting action of bandsawing differs from other sawing methods in that its continuous, single-direction cutting action, combined with blade guiding and tensioning, gives it the ability to follow a path that cannot be duplicated by power hacksawing or circular sawing. Band teeth cut with a shearing action and tend to take a full, uniform chip.

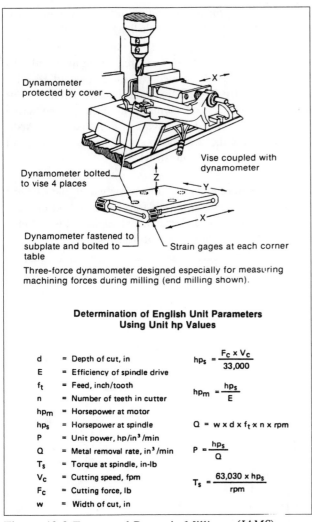

Three-force dynamometer designed especially for measuring machining forces during milling (end milling shown).

Determination of English Unit Parameters Using Unit hp Values

d	= Depth of cut, in	$hp_s = \dfrac{F_c \times V_c}{33,000}$
E	= Efficiency of spindle drive	
f_t	= Feed, inch/tooth	
n	= Number of teeth in cutter	$hp_m = \dfrac{hp_s}{E}$
hp_m	= Horsepower at motor	
hp_s	= Horsepower at spindle	$Q = w \times d \times f_t \times n \times rpm$
P	= Unit power, hp/in³/min	
Q	= Metal removal rate, in³/min	$P = \dfrac{hp_s}{Q}$
T_s	= Torque at spindle, in-lb	
V_c	= Cutting speed, fpm	
F_c	= Cutting force, lb	$T_s = \dfrac{63,030 \times hp_s}{rpm}$
w	= Width of cut, in	

Figure 19.9 Forces and Power in Milling. (*IAMS*)

Toothed bands with different tooth geometries and hardnesses for specific applications are used for conventional bandsawing tasks. Terminology generally accepted for saw bands is presented in Figure 19.10.

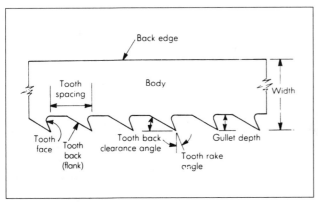

Figure 19.10 Bandsaw Terminology.

Three major types of tooth geometries, generally classified as standard, skip, and hook teeth, are illustrated in Figure 19.11.

Pitch, the number of teeth per inch, is determined primarily by the thickness of the material to be cut. General recommendations for band pitches to saw materials of different thickness are given in Table 19.1.

Other factors such as workpiece material and surface finish required must be taken into account in selecting the optimum pitch for a band. Optimum pitch is ensured if at least two teeth are in contact with the workpiece at all times during sawing.

Tooth set is the projection of the teeth from the sides of the band to provide cutting clearance and prevent binding. Overall set is the total distance between the outer corners of

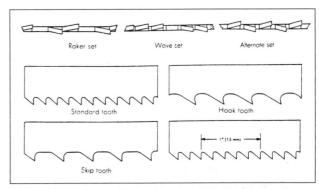

Figure 19.11 Three Major Types of Tooth Geometries.

Table 19.1
Recommended Band Pitches to Saw Materials of Different Thicknesses (*DoAll Co.*)

Material Thickness	Band Pitch
Less than 1" (25.4 mm)	10 or 14
1–3" (25.4–76.2 mm)	6 to 8
3–6" (76.2–152.4 mm)	4 to 6
6–12" (152.4–204.8 mm)	2 or 3
Over 12" (304.8 mm)	1 1/2 to 3

oppositely set teeth, which determines the kerf. The three most common types of set (Figure 19.11) are:

- Raker set,

- Wave set,

- Straight (alternate) set.

19.9 Grinding

Grinding lends itself to processes where high surface finish and high dimensional accuracy are needed. Grinding also can be a material removal process in which standard chip-type forming operations cannot cut harder materials. There are several types of grinding, including surface grinding, centerless grinding, and cylindrical grinding, among others.

Regardless of the type of grinding operation used, the identification system defined by ANSI standard B74.13-1977 is used by all grinding wheel manufacturers. This system uses letters or numbers in each of the seven positions, as detailed in Figure 19.12. Another ANSI standard covers the specifications of shapes and sizes of grinding wheels and mounted wheels.

In general, before mounting a grinding wheel, the ring test should be performed to check for cracks. Wheel dressers also are used to clean and true the wheel after it is mounted and also after certain periods of usage.

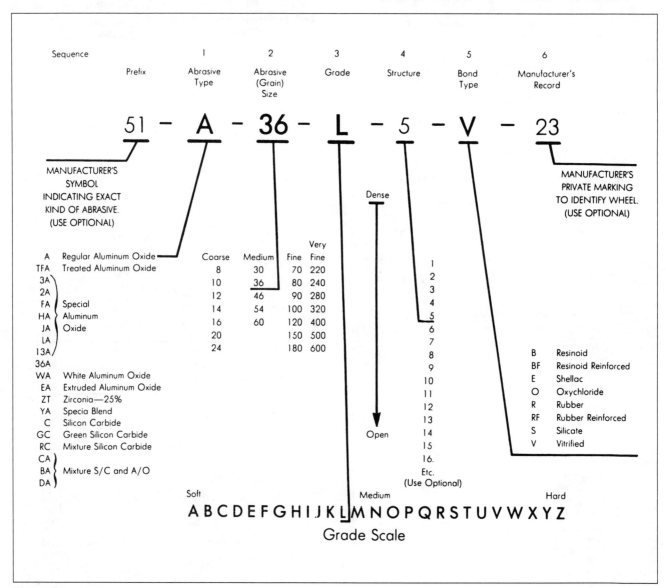

Figure 19.12 The System for Standards.

19.10 Punching/Blanking

When sheet metal is cut by blanking or punching, the force applied to the metal is basically a *shear force*. The total force is comprised of equal and opposite forces spaced a small distance apart on the metal. The capability of a material to resist this force is called *shear strength*. The shear strength of a material is directly proportional to its tensile strength and hardness. The shear strength increases as the tensile strength and hardness of a material increases. The blanking or punching force can be calculated by:

$$F = SLT$$

For round holes:

$$F = S \pi DT$$

where:

F = blanking or punching force, lb
S = shear strength of material, psi
L = sheared length, in.
D = diameter, in.
T = material thickness, in.

Example 19.10.1. Determine the force required to punch 3-in.-diameter blanks from 0.50-in.-thick steel plate when the shear strength of steel is 40,000 psi.

Solution. Force = 40,000 lb/in.^{2} × π × 3 in. × 0.50 in.
= 94.25 tons

The clearance between the punch and the die is very important. If the clearance is correct, fracture lines will form ideally at the punch and die. If the clearance is excessive, plastic deformation will occur because the punch will essentially pull the material into the clearance. The following formula can be used to calculate the clearance (one side):

$$c = at$$

where:
c = clearance (one side) in./mm
t = material thickness, in./mm
a = allowance %

Example 19.10.2. Determine the clearance for blanking 2-inch-round blanks in 0.250-in.-thick steel plate with a 5% allowance.

Solution. Clearance = 0.05 × 0.250 in.
= 0.0125 in.

19.11 Drawing

Drawing is a process of cold forming a flat precut metal blank into a hollow vessel without excessive wrinkling, thinning, or fracturing. The process of drawing basically involves forcing the flat sheet of metal into a die cavity with a punch. The force exerted by the punch must be sufficient to draw the metal over the edge of the die opening and into the die.

For drawing cylindrical shells having circular cross sections, the maximum drawing force may be calculated from the following equation:

$$P = n \pi dt\sigma_b$$

where:
P = drawing force, lb or kN
σ_b = tensile strength of the blank material, psi or MPa
d = punch diameter, in. or mm
t = sheet thickness, in. or mm
n = σ_D/σ_B

Example 19.11.1. Calculate the drawing force for a cylindrical shell from 0.250-in. steel, punch diameter of 2 in., n = 1.2, and tensile strength of 40,000 psi.

Solution. Force = 1.2 × π × 2 in. × 0.250 in. × 40,000 lb/in.^{2}
= 150.8 tons

19.12 Forging

Forging is defined as the controlled, plastic deformation or working of metals into predetermined shapes by means of pressure or impact blows, or a combination of both. Forging aligns the grain to the contour of the part, thus increasing its strength.

There are several types of forging processes such as open-die forging used to make simple shapes such as discs, rings, or shafts. In impression-die forging, the workpiece is placed between two dies containing the impression of the shape to be forged (Figure 19.13). Closed-die forging, or flashless forming, does not depend on flash to achieve complete die filling. The material is formed in a cavity that does not allow material to flow outside, which makes starting material volume calculations extremely critical.

An approximate estimate of the total deformation energy required for conventional forgings can be determined from the following formula:

$$E_D = V \times F_I \times S_F \times C$$

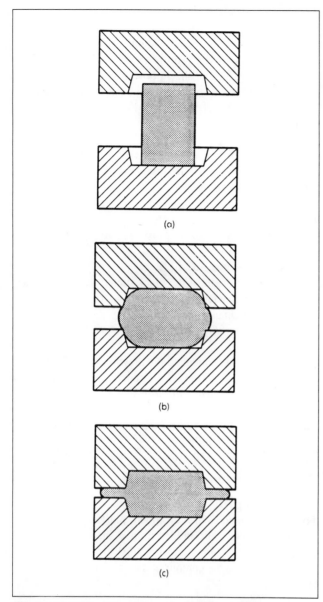

Figure 19.13 In impression-die forging, (a) The workpiece is inserted between the dies, (b) the dies are brought together deforming the workpiece until the sides come in contact with the walls, and (c) the thin flash assists the flow of material and completes die filling. (*American Foundrymen's Society*)

where:

E_D = total deformation energy required for finished forging, ft-lb

V = volume of billet to be forged, in.^{3}

F_I = forgeability index of material being forged

S_F = shape factor obtained from square root of ratio of average width and average thickness

C = constant equal to 2000

Example 19.12.1 Determine the total deformation energy in forging a 2-in.-diameter piece of inconel $\times$ 2 in. long. The forgeability index is 2.5 and the shape factor is 1.4.

Solution. $E_D = \pi \times 1^2 \times 2$ in. $\times 2.5 \times 1.4 \times 2000$
$= 43,982$ ft-lb

19.13 Energy Considerations in Press Work

The force rating of a mechanical press is a major consideration; however, energy (work) capacities are also important. Kinetic energy stored in the rotating flywheel of the press is the major source of energy to perform work. Useful energy in the flywheel can be calculated using the following equation:

$$E = N^2 \times D^2 \times W/\, 5.25 \times 10^9$$

For the metric version, replace 5.25×10^9 by 6.8×10^6

where:

E = available flywheel energy, in.-tons
N = rotary speed of flywheel, rpm
D = diameter of flywheel, in.
W = weight of flywheel, lb

Example 19.13.1. What is the available energy of a 500-lb flywheel with a diameter of 24 in. and turning at 200 rpm?

Solution. $E = 200^2 \times 24^2 \times 500$ lb$/5.25 \times 10^9$
$= 2.2$ in.-tons

For constant angular acceleration:

$$w = w_o + \alpha_c t$$
$$\Theta = \Theta_o + w_o t + .5\, \alpha_c t^2$$
$$w^2 = w_o^2 + 2\alpha_c\, (\Theta - \Theta_o)$$

where:

w = angular velocity
Θ = angular displacement
α = angular acceleration

Example 19.13.2. How long does it take for a flywheel starting from rest to reach 3500 rpm with a constant angular acceleration of 5 rad/sec^2?

3500 rev/min $\times$ 1/60sec $\times$ 2π = 366.52 rad/sec
t = 366.52 rad/sec/5 rad/sec^2
$= 73.3$ sec

19.14 Hot Extrusion

Extrusion is the plastic deformation process in which material is forced under pressure through one or more die orifices. In hot extrusion, heated billets are reduced in size and forced to flow through dies to form products of uniform cross sections, along their continuous lengths.

Extrusions are made using one of two methods: direct or indirect. With direct or forward extrusion (Figure 19.14), a billet of metal (No. 4) is placed in a heavy-walled container (Nos. 7, 8) and the extruded product (No. 1) exits through a die (No. 3) secured in a holder. The force for extruding is applied by a pressing ram (No. 6) with an intermediate, reusable dummy block (No. 5). Metal flow from the die is in the same direction as the forward movement of the ram.

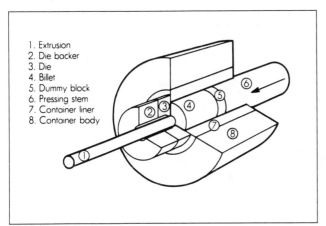

Figure 19.14 Direct Method of Hot Extrusion. (*Wean United*)

In indirect or backward extrusion, the billet remains stationary relative to the container wall while the die is pushed into the billet by a hollow ram; or the container and billet are pushed over the stationary ram (Figure 19.15). The die is loosely attached to the ram and the extrusion exits through the hollow ram.

For less complicated shapes such as round bars and tubes, pressure requirements can be calculated by using the following formula:

$P = A_o k ln(A_o/A_f)$
where:
P = extrusion pressure required, psi or MPa
k = extrusion constant
A_o = billet area
A_f = extruded product area

Example 19.14.1. A round billet of 1020 steel is extruded at 1900°F (1038°C). The billet diameter is 3 in. and the

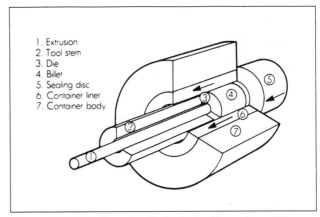

Figure 19.15 Indirect Method of Hot Extrusion. (*Wean United*)

extruded diameter is 2 in. Calculate the extrusion force required. Use 21 ksi for the k value.

Solution. $A_o = \pi\ 3^2/4 = 7.1$ in.^{2}
$A_f = \pi\ 2^2/4 = 3.14$ in.^{2}
$P = 7.1$ in.$^2 \times 21,000 \times ln\ (7.10/3.14)$
$= 121,600$ psi

19.15 Casting

Casting is a process in which molten metal is poured or injected into a cavity and allowed to solidify to the shape of the cavity. After solidification, the part is removed from the mold and then processed for delivery. Casting processes vary from simple to complex. Material and process selection depends on the part's complexity, function, quality specifications, and the projected cost level. Table 19.2 illustrates the range of materials for casting processes.

Figure 19.16 illustrates a typical green-sand mold section, showing the basic elements common to most casting processes. In most casting processes, the term to describe the molds is the same. Molds are usually, but not always, made in two halves. Exceptions are investment casting which uses a one-piece mold and die casting that can use dies made up of more than two parts for casting complex shapes.

Molds are generally made by surrounding a pattern with a mixture of granular refractory and binder mixture. This mixture can be wet or dry depending on the casting process. The choice of mold material depends on the casting quality and quantity as well as the type of metal to be used.

In most processes, the upper half of the mold is called the "cope" and the lower half is referred to as the "drag." Cores made of sand or metal are placed into the cavity to

Table 19.2
Commercial Capability of Casting Processes.

Process	Ductile Iron	Steel	Stainless Steel	Aluminum, Magnesium	Bronze, Brass	Gray Iron	Malleable Iron	Zinc, Lead
Die casting				•	•			•
Continuous	•				•	•		•
Investment	•	•	•	•	•			
Ceramic cope & drag	•	•	•	•	•	•		•
Permanent mold				•	•	•		•
Plaster mold				•	•			•
Centrifugal	•	•	•			•		•
Resin shell	•	•	•	•	•	•	•	•
Sand	•	•	•	•	•	•	•	•

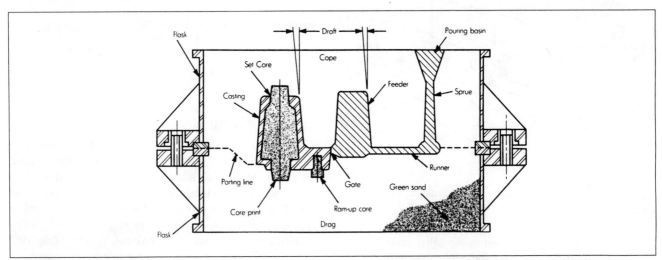

Figure 19.16 Cross-section View of Typical Cored Casting and the Sand Mold in Which It Is Produced. (*American Foundrymen's Society*)

form the inner surfaces of the casting. The mold requires a gating system to distribute metal in the mold and risers (liquid reservoirs) to feed the casting as it solidifies. The sprue is the channel, usually vertical, through which the metal enters. A runner, usually horizontal, leads the metal into the mold. A riser is a reservoir connected to the cavity to provide liquid metal to the casting to offset shrinkage as the casting solidifies. Finally, gates attach the entire gating system to the casting.

Castings generally exhibit nondirectional properties which make them weaker than wrought metals which are anisotropic-stronger and tougher in one direction vis-a-vis another. In some instances, the properties and performance attainable in cast components cannot readily be obtained by other manufacturing methods. For example:

- Castings allow the manufacture of parts from alloys that are difficult to machine or weld.

- Complex shapes are easier to produce by casting than by other processes.

- Parts with internal cavities also are easier and more economical to produce by casting than by other processes.

For simple shapes, near-net-shape casting often cannot compete economically with forgings. Castings are best used for complex part geometries and internal cavities.

19.16 Welding/Joining

As defined by the American Welding Society (AWS), there are five welding categories: (1) oxyfuel gas welding, (2) arc welding, (3) resistance welding, (4) solid-state welding, and (5) unique processes.

Oxyfuel Gas Welding. Oxyfuel gas welding generally uses combustion of oxygen and acetylene which burns at around 6300°F (3482°C) to provide heat. Known as oxyacetylene welding, it may be used with or without a filler material. The equipment used for oxyfuel welding, in general, is comparatively simple, compact, usually portable, and inexpensive, making it useful for maintenance and repairs.

Arc Welding. In arc welding, an electric arc extends from an electrode (consumable or nonconsumable) to the grounded workpiece. The electric arc generally creates welding temperatures of around 12,000°F (6649°C). Arc welding processes are numerous, the more popular of which are shielded metal arc welding (SMAW), gas metal arc welding (GMAW), and gas tungsten arc welding (GTAW).

SMAW (commonly referred to as stick-electrode welding) is an arc welding process that joins metals by heating them between a covered (or coated) electrode and the work (Figure 19.17). Shielding is provided by decomposition of the electrode covering. Pressure is not used, and filler metal is obtained from the electrode.

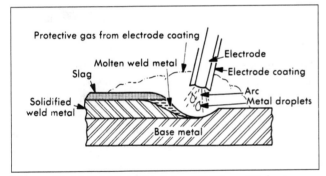

Figure 19.17 Shielded Metal Arc Welding. (*Hobart Institute of Welding Technology*)

GMAW (also known as MIG welding, dip transfer, or wire welding) is an arc welding process that joins metals by heating them with an electric arc between a continuous, solid (consumable) electrode for filler metal and the work (Figure 19.18). Shielding is provided by an externally supplied gas or gas mixture.

GTAW (also known as TIG welding or heliarc) is an arc welding process that produces joining of metals by heating them with an electric arc between a tungsten (nonconsum-

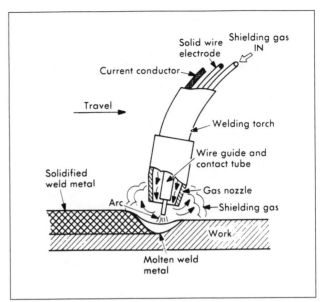

Figure 19.18 Gas Metal Arc Welding. (*Lincoln Electric*)

able) electrode and the work (Figure 19.19). Shielding is obtained by an envelope of an inert gas or gas mixture. Pressure and filler may be used.

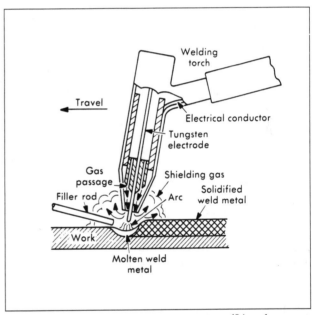

Figure 19.19 Gas Tungsten Arc Welding. (*Lincoln Electric*)

Resistance Welding. Resistance welding is a group of welding processes producing coalescence of metals with the heat obtained from the resistance of the work to electric current flowing in a circuit or which the work is a part, and by the application of pressure. No flux, filler metal, or shielding of any type is used. Although there are many resistance welding types, resistance spot welding is perhaps most common and the methodology applies to all resistance welding operations.

Resistance spot welding produces coalescence at the faying surfaces in one spot. The size and shape of the individually formed welds (called nuggets) are influenced primarily by the size and contour of the electrodes. Most spot welding is done by clamping the workpieces between a pair of electrodes and passing a low-voltage, high-amperage current through the electrodes and workpieces for a short cycle. Resistance heating at the joint contacting surfaces forms a fused nugget of weld material. The heat in resistance welding obeys the following equation:

$$H = I^2RT$$

where:

H = heat
I = current
R = resistance
T = time

The resistance is determined by:

- Resistance of the electrodes,

- Resistance of the workpieces,

- Distance between the faying surfaces, and

- Electrode pressure.

Example 19.16.1. Determine the energy absorbed for the following resistance welding operation: 50 amps, 2 seconds, and 0.20Ω.

Solution. $H = 50^2 \times 0.20\Omega \times 2$ seconds
$= 1000$ Joules

Brazing and Soldering. Brazing is defined by the American Welding Society (AWS) as "A group of welding processes which produces coalescence of materials by heating them in the presence of a filler metal having a liquidus above 840°F (450°C) and below the solidus of the base metal. Liquidus is the lowest temperature at which a metal or alloy is completely liquid, and solidus is the highest temperature at which a metal or alloy is completely solid." The filler metal is distributed between the closely fitted faying surface of the joint by capillary action. Braze welding differs from brazing in that the filler metal is not distributed in the joint by capillary action. In brazing, flux generally prevents oxidation and promotes wetting, flow, and the formation of a soundly brazed joint.

Soldering. Soldering is defined by the American Welding Society as "A group of welding processes which produces coalescence of materials by heating them to a suitable temperature and using a filler metal having a liquidus not exceeding 840°F (450°C) and below the solidus of the base material. The filler metal is distributed between the closely fitted surfaces by capillary attraction."

19.17 Spray Finishing

In conventional air spray, the material is usually supplied from a container in one of two ways. The container may be under pressure, up to 100 psi (690kPa) or the spraying device can pull material from the container to the atomizing area (suction feed).

A typical air-atomizing system consists of: (1) air pressure source, (2) air regulator, (3) air line, (4) material supply, and (5) spray device.

A type of spray gun that controls only the fluid flow is known as a bleeder-type because the air constantly bleeds from the gun as it is being used. The other type of spray gun, known as the nonbleeder-type, controls the air and the fluid by the action of the trigger. These guns ensure that the air comes on before the fluid begins to flow, known as lead-lag.

The air nozzles, referred to as air caps, are the most important part of the air spray gun. They direct the air to the material and cause atomization and pattern development. The two basic types of air nozzles are external-mixing and internal-mixing. External-mixing systems (Figure 19.20) mix the air and the fluid outside the air cap. This type of cap is used on both bleeder and nonbleeder types of spray guns and can be either siphon- or pressure-fed. External-mixing systems are the most common type used in production.

Internal-mixing systems (Figure 19.21) mix the air and the fluid inside the air cap before being released. The air cap's exit-hole shape controls the pattern of the material spray, which cannot be varied with the gun controls. Internal-mixing systems must be pressure fed and the air and fluid balance must be closely maintained.

19.18 Electrostatic Spraying

In this finishing process, the application of electrostatic charges to the material particles causes them to act like small magnets when placed in the vicinity of a grounded

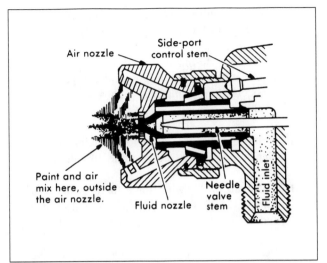

Figure 19.20 External-mixing System Spray Gun. (*Binks Manufacturing Co.*)

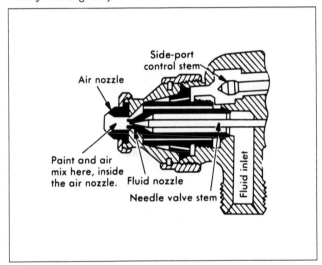

Figure 19.21 Internal-mixing System Spray Gun. (*Binks Manufacturing Co.*)

object. During the spraying process, the part to be painted is grounded. As the material is sprayed toward the part, the magnetic action of the charged particles causes the particles, normally lost due to bounce-back or blow-by, to be attracted back to the part by actual wrapping of the part in material particles. This phenomenon is known as "wrap" and is the prime force in the move to electrostatics. By applying an electrostatic charge to the material particles, transfer efficiencies of 60% to 90% are possible.

19.19 **Plating**

In electroplating, the workpiece is made cathodic in a solution containing the ions of the metal being deposited. Direct current is passed between the anode and the workpiece (cathode). The anode is usually constructed of the same material as the metal being plated. As the current flows, the metal ions gain electrons at the cathodic workpiece and transform into a metal coating.

Proprietary additives are usually incorporated in the plating solution to alter the deposit in a desirable fashion. These additives brighten or level the deposit as well as improve the uniformity of the deposit's thickness over the entire workpiece. The additives also may be used to alter physical properties such as hardness, ductility, internal stress, and corrosion resistance.

Many different metals can be successfully electroplated, such as nickel, copper, chromium, zinc, tin, cadmium, and lead. Alloys and precious metals also can be deposited.

19.20 **Jigs/Fixtures/Locating**

A jig is a device locating and holding a workpiece while guiding or controlling a cutting tool. A fixture is simply a locating and holding device having nothing to do with tool guidance or control.

In good design, the fixture must confine the workpiece through six degrees of freedom: three linear and three rotational. This idea leads to the 3-2-1 principle, which states that a workpiece will be confined when placed against three points in one plane, two points in another plane, and one point in a third plane, if the planes are perpendicular to each other.

Generally, locating devices can be either a machined surface on the fixture to support the workpiece, pins aligned to locate finished holes, or vee blocks. Typically, rest buttons are used in place of machining a pad on the fixture. These buttons have hardened heads and tempered shafts. Locating pins can sometimes cause part jamming. To reduce the amount of jamming, several parameters can be changed, such as the length of fit and an alignment groove. Cutting three equal flats on the length of the pin (diamond locator) also reduces jamming.

There are many guidelines for designing or choosing clamps. Clamps come in several types, including screw, cam, level, toggle, wedge, and latch. In general, clamps should hold the workpiece against a locating surface. The clamping force should be transmitted through a fixed support point on the workpiece in a manner that does not distort the workpiece. Fixed locators should oppose the force of the cutting tool, not the clamps. Clamping forces can vary based on depth of cut, dulling of cutting tools, change of material hardness, and cutting rate changes.

Practice Problems

19.1 Find the rating factor if the normal time is 15 seconds and the average assembly time is 20 seconds.

19.2 Calculate the standard time if the normal time is 10 seconds and an allowance of 10% is given due to assembly fatigue.

19.3 Determine the shear angle, ϕ, if the cutting ratio is 0.25 and the rake angle is 8 degrees.

19.4 What rpm should be used for turning a piece of low-carbon steel of 1-in. diameter with a cutting speed of 100 fpm?

19.5 Calculate the maximum depth of cut for a 3-in.-round 1020 steel on a lathe with a 3-hp motor, a feed rate of 0.0052 in./rev, 100 fpm cutting speed, 0.80 efficiency, and unit horsepower of 1.

19.6 Calculate the cutting speed of a material if a 0.50-in. drill is rotating at 1500 rpm.

19.7 What are the spindle horsepower requirements for drilling a 1-in. hole in aluminum with 300 fpm cutting speed, 0.00078 in./rev feed rate, 75% efficiency, and a unit horsepower of 1?

19.8 Determine the feed rate per tooth of a 4-in. milling cutter with 20 teeth, cutting aluminum using 300 fpm for a cutting speed, and a feed rate of 4 ipm.

19.9 Calculate the motor horsepower if a 2-in. by 2-in. cutter with 15 teeth is making a 0.375-in.-deep cut at 120 rpm, with a feed rate of 0.00055 in./tooth, unit horsepower of 1, and 75% efficiency.

19.10 Determine the maximum thickness of material that can be punched with a 2-in.-diameter punch, 50-ton press, working a material with a shear strength of 35,000 psi.

19.11 Determine the clearance for blanking 3-in.-square blanks in 0.500-in. steel with a 10% allowance.

19.12 Calculate the drawing force for a cylindrical shell from 0.125-in.-thick steel, punch diameter of 2.5 in. and tensile strength of 40,000 psi. n = 1.5.

19.13 Determine the total deformation energy in forging a 4-in.-diameter piece of steel 2 in. long with a forgeability index of 1.5 and a shape factor of 1.

19.14 Determine the weight of a 30-in.-diameter press flywheel if 1.5 in.-tons of available energy is needed at 150 rpm.

19.15 What angular acceleration is needed for a flywheel to reach 2000 rpm in 10 seconds starting from rest?

19.16 A round billet of 1020 steel is extruded at 1850°F (1010°C). The billet's original diameter is 2 in. and its final diameter is 1.780 in. Calculate the extrusion force required using 25 ksi for a k value.

19.17 Determine the energy absorbed for resistance welding steel at 100 amps, a resistance of 0.15Ω for 3 seconds.

MANAGEMENT AND ECONOMICS

20.1 Ethics in Management

Ethics refers to the system of moral principles and values established and/or demonstrated by individuals.

Ethics is neither very precise nor organized in business, despite the entire field of business law and a ponderous amount of business literature written on relevant moral issues. A high standard of human behavior and the good principles guiding it are the core of civilized life. Further, many times, those ethical people who follow good principles are trusted, respected, honored, and favorably viewed by people in general and especially by one's business associates.

From a more individual standpoint, a person who is unethical simply cannot be tolerated by a business organization whether the function is at the top or bottom of a firm. There are a great many money-related opportunities in commerce tempting those with weak principles. Therefore, good conduct must be immediate and instinctive.

Reasons for Importance. Because so much is contractual and promissory in nature, the ethical qualities of participants are central. A business must operate through agents and get its job done through people, so this concern extends from the top of the organization, which sets overall policies, to the bottom, which follows the example set by management. Management should: (a) demonstrate and demand ethical behavior, (b) take clear and swift action to wrong conduct, and (c) place value on ethics in training and performance evaluations.

For example, suppose a flaw exists in a minor part used in a jetliner. This flaw, and its use in the airplane, can result in an extremely costly loss to the fleet, and in human lives and property, if a crash results. Insurance rates would climb, lawsuits mount, and the airline and airframe manufacturer may experience such a crushing loss in reputation that their futures may be in jeopardy. Of course, they go on

to sue the parts manufacturer, which would probably have to close its doors. And all of this can occur if a single worker chooses to act unethically and knowingly uses a bad part rather than be delayed in his or her production activity. The possibility of this happening can be reduced if management sets and expects a high standard and example of ethical behavior.

Codes. An ethical problem in business is often a gray area beyond the reach of existing laws but which may, at some future date, be so formulated. One tool for dealing with gray area matters is for each company, industry, and profession to develop a code of work behavior guidelines. Of course, where employees must abide by both professional and company codes, there can be an overlap so long as they are not in conflict.

Professional codes may serve two purposes. They can recommend appropriate rulings to the firm as well as help shield the professional from having to carry out inadvisable actions. A single firm might have to deal with codes of varying content and strength from a number of outside societies that can be local, regional, or national in authority. This is similar to dealing with different unions in which every employee involved is not a member of the authoring body. Further, international firms can experience greater complications as slightly different ideologies try to merge into one. Other complicating factors come into play as well, such as the various cultures, values, and ethical perspectives of countries other than the U.S.

Approximately 75% of U.S. firms have a written code of ethics. It should be recognized that differing conditions in separate firms make it unrealistic to standardize a detailed code of ethics for all organizations. A code of ethics is somewhat specific to an organization, as it aids in defining and guiding real-life practices. A professional code can aid in conformity, however. SME's code, Figure 20.1, adopted in April 1975, is a good example of a professional code of ethics.

SOCIETY OF MANUFACTURING ENGINEERS

CODE OF ETHICS

Preamble

Practitioners of manufacturing engineering recognize that their professional, civic and personal activities have a direct and vital influence on the quality of life and standard of living for all people. Therefore, manufacturing engineers should exhibit high standards of competency, honesty and impartiality; be fair and equitable; and accept a personal responsibility for adherence to applicable laws, the protection of the public health, and maintenance of safety in their professional actions and behavior. These principles govern professional conduct in serving the interests of the public, clients, employers, colleagues and the profession. Honesty, integrity, loyalty, fairness, impartiality, candor, fidelity to trust, and inviolability of confidence are incumbent upon every member as professional obligations. Each member shall be guided by high standards of business ethics, personal honor, and professional conduct. The words "practitioner," "manufacturing engineer," and "member" as used throughout this Code include all classes of membership in the Society of Manufacturing Engineers.

The Fundamental Principle

The manufacturing engineer is dedicated to improving not only the manufacturing processes, but manufacturing enterprises worldwide. This includes striving to instill a sense of concern and awareness throughout the manufacturing community of public health, safety, conservation, and environmental issues that are related to the practice of manufacturing and through the application of sound engineering and management principles. Engineers realize that in carrying out this responsibility their individual talent and services can be more effective when funneled through the activities of the Society of Manufacturing Engineers. Therefore, engineers shall strive to support the mission of the Society of Manufacturing Engineers and the activities, products, and events sponsored and produced by them.

Canons of Professional Conduct

Members offer services in the areas of their competence and experience, affording full disclosure of their qualifications.

Members consider the consequences of their work and societal issues pertinent to it and seek to extend public understanding of those relationships.

Members are honest, truthful, and fair in presenting information and in making public statements reflecting on professional matters and their professional role.

Members engage in professional relationships without bias because of race, religion, sex, age, national origin or impairment.

Members act in professional matters for each employer or client as faithful agents or trustees, disclosing nothing of a proprietary nature concerning the business affairs or technical processes of any present or former client or employer without specific consent.

Members disclose to affected parties known or potential conflicts of interest or other circumstances which might influence—or appear to influence—judgment or impair the fairness or quality of their performance.

Members are personally responsible for enhancing their own professional competence throughout their careers and for encouraging similar actions by their colleagues.

Members accept responsibility for their actions; seek and acknowledge constructive criticism of their work; offer honest constructive criticism of the work of others; properly credit the contributions of others; and do not accept credit for work not theirs.

Members perceiving a consequence of their professional duties to adversely affect the present or future public health and safety shall formally advise their employers or clients and, if warranted, consider further disclosure.

Members of the Society of Manufacturing Engineers act in accordance with all applicable laws and the Constitution & Bylaws of the Society of Manufacturing Engineers and lend support to others who strive to do likewise.

Members of the Society of Manufacturing Engineers shall aid in preventing the election to membership of those who are unqualified or do not meet the standards set forth in this Code of Ethics.

Approved by: Society of Manufacturing Engineers Board of Directors
Date: December 2, 1990

Figure 20.1 Professional Code of Ethics Adopted by the Society of Manufacturing Engineers.

20.2 Economics

Value Engineering. Value engineering provides a systematic approach to evaluating design alternatives. Value engineering is often very useful and may even point the way to innovative new design approaches or ideas. It is also called value analysis, value control, and value management. In value engineering, a multidisciplinary team analyzes the functions provided by the product and the cost of each function. Based on the analysis results, creative ways are sought to eliminate unnecessary features and functions and to achieve required functions at the lowest cost while optimizing manufacturability, quality, and delivery.

In value engineering, "value" is defined as a numerical ratio, the ratio of function (or performance) to the cost. Because cost is a measure of effort, value of a product using this definition is simply the ratio of output (function or performance) to input (cost) commonly used in engineering studies. In a complicated product design or system, every component contributes both to the cost and the performance of the entire system. The ratio of performance to cost of each component indicates the relative value of individual components. Obtaining the maximum performance per unit cost is the basic objective of value analysis.

Functional and Esteem Value. For any expenditure or cost, two kinds of value are received: functional (use) value and esteem (prestige) value. Functional (use) value reflects the properties or qualities of a product or system that accomplish the intended work or service. To achieve maximum functional value is to achieve the lowest possible cost in providing the performance function. Esteem value is composed of properties, features, or attractiveness making ownership of the product desirable. To achieve maximum esteem value is to achieve the lowest possible cost in providing the necessary appearance, attractiveness, and features that the company wants. Examples of prestige items include surface finish, streamlining, packaging, decorative trim, ornamentation, attachments, special features, and adjustments.

Waste. In addition to the two kinds of value received, additional costs come from unnecessary aspects of the design. Termed "waste," these are features or properties of the design providing neither use value nor prestige value.

Typical scales of value for some common, well-known items are shown in Figure 20.2.

Tie	5% function	90% esteem		5% waste
Hammer	80% function		15% esteem	5% waste
Tie clasp	20% function	75% esteem		5% waste
Button	90% function		10% esteem	

Figure 20.2 Estimated Scale of Value of Some Common Products. (*Industrial Technology Institute*)

Production Planning and Control. Essential to any manufacturing support activity is production planning and control. Often, the principal group of personnel involved in this process is called production and inventory control, which may be part of a materials management or manufacturing management department. Production planning and control involves forecasting, aggregate planning and master

scheduling, materials requirements planning (MRP) or manufacturing resource planning (MRP II), process planning, inventory management, and systems integration, among others.

Forecasting. The production and inventory planning process begins with forecasting. Business operations require some technique of predicting future demand for business planning and personnel decisions. The anticipated demands and the forecast error rates experienced determine the basis for the systems, inventory policies, purchasing practices, and shop scheduling techniques supporting production requirements. Four key concepts and principles describe forecasting in this context:

1. Forecast accuracy is indirectly proportional to the length of time of the forecasted period; the shorter the forecast period, the more accurate the forecast.

2. Forecast accuracy is directly proportional to the number of items in the forecast group. The total company forecast can be expected to be more accurate than the corresponding forecast for a given product line, which in turn will be more accurate than the corresponding forecast for a single part number in that product line.

3. Forecast error is always present and should be estimated and measured on all forecasts.

4. No single forecast method is best; alternative methods should be tested periodically to determine if another method would result in a smaller forecast error.

Approaches used are: (a) a study of seasonality based on historical data to support a calculation, (b) a study of trends using least-square regression, (c) a study of residual variation not explained by the model to determine if the variation is random or due to special causes.

Aggregate Planning and Master Scheduling. Aggregate planning by top-level management is based on forecasts. This planning describes the framework for developing a detailed master schedule and production plan by establishing a sales plan, inventory plan, and production plan as part of the aggregate plan. Aggregate planning is also known as PSI (production, sales, inventory) planning.

The sales plan begins with a shipments forecast and considers production management concerns for flexibility and factory capability. The inventory plan allows a strategic approach to investment in inventory and inventory levels rather than those levels being set as a reaction to detail-level policies and procedures. A production plan is the final part of aggregate planning by top-level management. It is the starting point from which a master schedule and detailed production plan is developed.

With the completion of the aggregate plan, top-level management authorizes the production planning department to develop a master production schedule (MPS). The MPS is the primary document of a factory plan of operation for a period of time equal to its longest lead times. It also serves as the allocation vehicle of expected or confirmed orders to time slots for production. The master production schedule is the actual top level of input into an MRP system.

Requirements and Capacity Planning. From the master production schedule, a detailed production plan must be developed. This plan focuses on production and material control and is intended to balance limited resources with production planned on the master schedule. The detailed production plan must consider the economics and feasibility of the alternatives as they relate to inventory investment, storage capacity, purchased component availability, and personnel availability.

Process Planning. Once a part is designed, the production processes must be planned. Process planning involves the development of operation flow charts, production layouts, routings, operation (process) sheets, setup charts and machine tool layouts, equipment selection and sequence, material handling details, tooling requirements, inspection plans for quality assurance and quality control, production cost analysis, and much more.

Computer-aided process planning (CAPP) systems are expert computer systems that collect and store the knowledge of a specific manufacturing situation, as well as general manufacturing engineering principles. This information is used to create the optimum plan for manufacturing a new part. This plan specifies the machinery to be used for production, the sequence of operations, the tooling, required speeds and feeds, and other necessary data.

With CAPP, the part is designed on a CAD/CAM system. The part file is transferred into the CAPP system, matching the part characteristics to the machines and processes available on the shop floor. The CAPP system then prints out the process and routing sheets making up the process plan.

Two main types of CAPP systems—variant and generative—are in use today. The variant system modifies the process plan for a similar, previously produced part to produce a plan for the new part. The generative system starts from scratch when developing a process plan, and

therefore needs a large database containing manufacturing logic, capabilities of existing machinery, standards, and specifications.

Scheduling and Production Control. The priorities of the sequence of jobs to be run in the shop must be established:

1. A priority system should specify the jobs to be done first, second, third, and so on.

2. A priority system should allow for easy and fast updating of priorities, as priorities change after a very short time because actual conditions change.

3. A priority system should be objective. If jobs are overstated, an "informal" system determines which jobs are really needed.

Some of the commonly used priority schemes in use today are: first in/first out, start date, due date, critical ratio rule, slack time ratio, and queue ratio.

Based on the prioritized schedule, work is authorized and the shop floor receives a shop order. Production and control of each shop order is tracked by collecting and analyzing the data from the workcenters through which each shop order must pass. The goal is to identify any production problems as soon as possible so that action can be taken to correct the problem and get production back on schedule.

Inventory Management. Inventory is one of the most important financial assets present in manufacturing companies. Stocks of raw materials, work-in-process inventory, and finished goods constitute the focus of control for the time they are held before being converted into sales dollars. The shorter the period that inventory is held, the more productive the asset.

Item demand is considered to be *independent* when that demand is unrelated to the demand for other items. Product demand arriving through orders is the principal element of this type of demand. This form of demand is usually associated with the manufacturer's primary revenue source and is the subject of most of the forecasting effort. Demand for items that will be consumed in destructive testing and service parts requirements are likewise independent demands.

Demands for parts or raw materials are considered to be *dependent* demands when they are derived directly from the demands for other items. The usual source of these requirements is the output of a bill of material "explosion." These demands are then accumulated as component and material requirements by time period. Such demands are therefore calculated and should not be forecasted independently. Some items are subject to both independent and dependent demands.

Demand for products, parts, components, and materials is rarely stable over time. Forecasting addresses the trends and variations of demand, but since forecasting is an estimate of demand, inventory management must accommodate forecasting error.

Inventory is replenished according to some set of rules, either formal or informal. The objective of inventory policies must be to balance the cost of carrying inventory with the service level required. The principal measure related to this activity is called inventory "turns." The equation for calculating inventory turn ratio is:

$$\text{Inventory turns} = \frac{\text{Annual inventory usage \$ at cost}}{\text{Average inventory \$ at cost}}$$

Conventional/traditional manufacturers typically experience turn ratios in the range of 2 to 10, while companies using just-in-time (JIT) techniques have ratios of 10 to 50 or more. Stock in inventory costs money, so more companies are using JIT to have as little in inventory as possible, and to achieve as high a turn ratio as possible. The use of "safety stock" to reduce the risk of a stockout will lower the turn ratio. Companies using JIT also reduce or eliminate safety stock.

Economic order quantity (EOQ) has been the most common use of a statistical calculation in inventory control for several decades. The standard EOQ formula is based on the model shown in Figure 20.3.

EOQ attempts to balance inventory carrying costs with the ordering costs. Annual usage in pieces is required as the first estimate in the calculation. The approximation of ordering costs must include setup costs if the part is a manufactured item rather than a purchased one. Inventory carrying costs result from the multiplication of the cost of one item by the management policy variable. This last factor describes the interest rate percentage believed to be a forecast of appropriate costs, including the cost of money, the cost of storage, the cost of handling, the cost of storage loss, and the costs associated with inventory obsolescence. While the calculation of EOQ is an estimate based upon estimates, it is useful in preliminary analyses. The formula for EOQ is as follows:

$$EOQ = (2AS)/ic$$

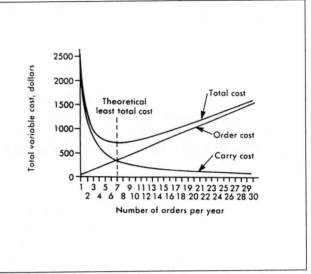

Figure 20.3 Order Cost Versus Inventory Carry Cost Curve. (*K.W. Tunnell Company, Inc.*)

where:
EOQ = economic order quantity
 A = annual usage
 S = setup and order costs per order
 i = interest and storage costs percentage
 c = unit cost of one part

Example 20.2.1 Calculate the EOQ for a product that has an annual usage of 20,000, setup and order costs of $50, and an interest and storage cost of 12%.

Solution. Inserting the figures in the example into the formula given results in an EOQ of 913.

A popular technique that lends itself to good management of the inventory asset dollar is the classical "ABC" analysis, which results in the coding of items by categories called A, B, and C. This technique requires sorting of items by the amount of dollar demand (at cost) recorded over some past period or from the output of an MRP system projected over some future period. It is based on Pareto analysis. It is usually observed that only about 20% of the items in any inventory will be involved in 80% of the usage measured by dollars. If this top 20% is managed carefully, the lower dollar items can be handled less often with little effect on the total dollar investment. Therefore, it is appropriate to base inventory policy statements on the basis of ABC analysis as a method of establishing an inventory plan. Basing inventory policies on the ABC analysis results in items being given replenishment rules like the following:

1. Review A items weekly, and order one week's supply when less than a lead time plus one week's supply remains.

- Review B items biweekly, and order four weeks' supply when less than a lead time plus two weeks' supply remains.

- Review C items monthly and order 12 weeks' supply when less than a lead time plus 3 weeks' supply remains.

The result of such policies and procedures is that the high dollar volume items get the most attention. In this example, the A items will be individually reviewed 4 times as frequently as the B items and 12 times as frequently as the C items.

Material Requirements Planning (MRP). The planning and control systems techniques based on computer-aided manipulation of data in time-phased "buckets" are now about 25 years old. It has become common practice to refer to such systems as some form of an MRP system. The introduction of JIT principles to U.S. manufacturing during the past few years created a different set of needs for such tools than those in use prior to 1980.

The typical material requirements plan is based on time-phased planning, and it originates from the components requirements planning needs of a manufacturing organization.

It is important to recognize that MRP is not for all businesses. There are single-product firms with simple bills of material that do not require MRP for manufacturing management activities related to planning and scheduling.

MRP is not just for replenishing inventories. The past decade saw the focus shift from simpler material requirements planning systems to total integrated manufacturing control systems, such as manufacturing resource planning (MRP II), discussed in a following section.

MRP is an exception-message, action-oriented system. It is not a clerically burdened system, unless it is improperly implemented. The discipline required in the management of this process becomes one of managing the variables' data associated with the control process. MRP techniques provide a method for routine calculation of:

- The "when" (timing of an order release),

- The "how much" (quantity required for order release), and

- "Priorities" (continually updated, based on purchasing and shop activities).

Properly used, the systems approach provides shop loading data to help make judgements about the uses of available capacity. It allows for the varying of demands and/or rules and/or lead times as a means of simulating results by asking "what if" questions.

MRP is a set of procedures, a set of decision rules and policies that govern many of the routine decisions required in setting the manufacturing schedule. As such, it provides a highly disciplined approach for arranging lower-level factory schedules. Its exception-action orientation is not a clerical system in nature, although when out of control, an MRP system can become a tremendous clerical burden. One critical aspect of the definition is that it is a highly disciplined management process. MRP depends on shop events happening just as they were simulated by the computer, based on the plans entered as the master production schedule (MPS) and the policies and operations data loaded in their databases.

Given the rules and procedures implemented for a given company, MRP determines the time and the quantity for order releases at lower levels and part manufacture requirements supporting the finished product schedules. MRP is not for controlling finished goods. MRP systems are driven by a master scheduling approach which is a finished product scheduling technique that must be developed in some manner to create the top-level demands that will be supported by MRP's subsequent arithmetic calculations.

The first calculations explode the demands into requirements by time period for lower-level subassemblies and component requirements. Demands are netted against available orders and committed replenishments (both shop orders and purchase orders) to calculate the actions required to support the master schedule. Two types of messages with associated quantity calculations result: (a) new order requirements and (b) changes required to existing orders. Order change messages may indicate data change needs only, quantity change requirements, or both.

The basis for improving operations through an MRP system requires that the functions related to scheduling be integrated and that it be driven by a valid MPS. Those who operate and manage the variables of the MRP system also must be qualified, which means educated and trained in not only the techniques of a given system, but also in the principles they are dealing with.

Procedures and controls for data accuracy are primarily operating discipline issues. The single largest failing in most MRP systems is the lack of discipline in the day-to-day activities that maintain data integrity within the system. A system attempting to emulate the total production environment within the computer depends on accurate data in the following elements:

1. Inventory balances,

2. Bills of materials,

3. Process routings,

4. Shop-order status by operation, and

5. Purchase order status by item and date.

MRP does not, in and of itself, reduce inventory or cause a reduction in inventory, nor can it improve customer service or productivity or reduce costs. As a tool, however, it can provide the means for management to gain those benefits.

The elements of an MRP database are shown in Figure 20.4.

Manufacturing Resource Planning (MRP II). Manufacturing resource planning is both logic for gaining manufacturing control and computer-based application systems employed to support this logic. The MRP II conceptual model shown in Figure 20.5 outlines the sequence of activities and decisions needed to provide a framework for planning, executing, controlling, and replanning production activities.

Market demands for products, service, or repairs are organized in quantitative terms useful to manufacturing. The demands include customer orders, changes desired in final stock available for sale, distribution inventories, and customer demand forecasts.

When properly organized, market demand portrays a production facility scheduled over relatively long periods of time, such as 6 to 18 months into the future. This is accomplished in production planning. The model assumes this decision is an approximate result representing management's business objectives that manufacturing needs to satisfy.

The production plan is examined in terms of aggregate capacities of the producing facility by reviewing overloads and underloads at major producing workcenters. The master planner (a person) iterates the placement of production requirements over time until a manageable and acceptable result is produced.

Master production scheduling is the decision process for placing the production of saleable products or models with options, for example, with respect to time. The critical

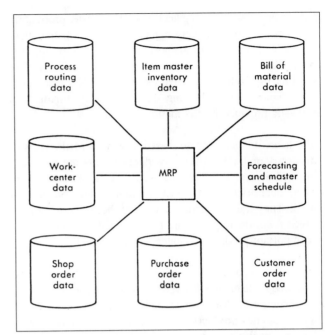

Figure 20.4 Diagram of Elements in an MRP Database. (*K.W. Tunnell Company, Inc.*)

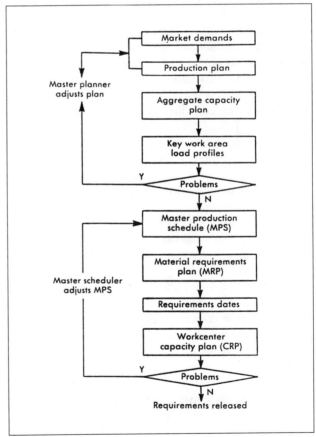

Figure 20.5 MRP II Conceptual Model. (*Coopers & Lybrand*)

time is the manufacturing lead time. The master production schedule (MPS) is usually regenerated for each manufacturing lead time within the horizon time of the production plan. Here the computer systems help examine the relative stability of production activities over the MPS lead time. The "smoothing"—looking through manufacturing lead time—is provided by the master scheduler.

The heart of the computer-based system is MRP. Bills of materials or parts lists for products/models are exploded into requirements for purchases, fabrications, and so on. Their quantities and time of availability are forecasts of need. They are calculated to support the execution of the MPS. Also, the requirements for labor and/or machine time at important production centers are calculated.

Next, the capacities of individual producing centers are analyzed to determine if large overloads are projected in the short run for the MPS-generated load. The master scheduler then simulates production activity until a reasonable work level is available at planned workcenters on a day-to-day basis.

A series of control feedback activities ensures that work is released to producing centers in concert with the MPS and MRP. Shop activity is relayed to the master scheduler in terms of whether or not production is ahead of or behind the master production schedule (variance control). Routine adjustments to MRP and released work (jobs, runs, rates per product) are made to recognize real production achievement, thus completing the control system process.

The production planner forwards scheduled plant activity to meet manufacturing's policies for smoothing overloads/ underloads in some sense. Master production scheduling disaggregates these results through the master scheduler.

Beginning with MRP, the computer system performs a series of necessary calculations. The master schedule is converted into a forecast of components, fabrications, and/or subassemblies using a standard product structure of ingredients related to MPS items. The quantities of purchased parts and manufactured parts/subassemblies are determined with respect to time. Quantity rules for satisfying requirements include lot/run for run, replenishment, or lot-sizing algorithms.

Commitments for purchases, manufactures, and their timing are determined using standard lead times to acquire (produce) components. This is a back-scheduling technique. Back-scheduling requires availability of purchased parts and the result of fabrication to match the dates determined by the component lead times.

The entire process provides the human scheduler with feedback concerning the practicality of executing the MPS.

20.3 Group Technology

Group technology (GT) is an approach to design and manufacturing to reduce manufacturing system information content by identifying and exploiting the sameness or similarity of parts based on their geometrical shape and/or similarities in their production process. GT is implemented by using classification and coding systems to identify and understand part similarities and to establish parameters for action. Manufacturing engineers can decide on more efficient ways to increase system flexibility by streamlining information flow, reducing setup time and floor space requirements, and standardizing procedures for batch-type production. Design engineers can develop an attitude for producibility and help eliminate tooling duplication and redundancy.

Group Technology and Design. Group technology can be applied in a variety of ways to produce significant design efficiency and product performance and quality improvements. One is its use to facilitate significant reductions in design time and effort. In design, it is often erroneously deemed easier to design new parts, tooling, and jigs rather than try to locate similarly designed parts. The ease with which parts can be designed using CAD systems exacerbates the problem. A GT database helps reverse this tendency by enabling the quick and easy retrieval and review of existing parts similar to the new part being designed. With GT, the design engineer needs only to identify the code describing the desired part. A search of the GT database reveals whether a similar part exists. If a similar part is found—and this is most often the case—the designer can simply modify the existing design to design the new part. In essence, GT enables the designer to start the design process with a nearly complete design. Group technology also facilitates standardization and rationalization (S&R), which helps control part proliferation and eliminates redundant part designs.

For example, a designer may find a gear identical to the one being designed but of a different thickness. Simply copying the existing design and making minor changes saves substantial design time and effort by helping to prevent redesigning the wheel.

In another case, it is common for a company to have many similar versions of the same part, such as a gear. When the company implements GT, similarities among gears can be identified, and it is possible to create standardized gears that are interchanged in a variety of applications and products. S&R such as this pays big dividends in that it simultaneously creates both economies of scale by increasing part volume and economies of scope because the same gear can be used in a variety of applications.

Part Families. The grouping of related parts into part families is the key to group technology implementation. The family of parts concept not only provides the information necessary to design individual parts in an incremental or modular manner, but also provides information for rationalizing process planning and forming machine groups or cells that process the designated part family. A part family may be defined as a group of related parts possessing some specific sameness and similarities. Design-oriented part families have similar design features, such as geometric shape. Manufacturing-oriented part families can be based on any number of different considerations. Such considerations may include parts manufactured by the same plant, parts that serve similar functions such as shafts or gears, or parts fabricated from the same material. All these parts could conceivably be grouped into part families.

Grouping Methods. Three methods of grouping parts are commonly used: visual inspection, production flow analysis (PFA), and classification and coding. Visual inspection of parts and their drawings is quite simple but limited in effectiveness when a large number of parts is involved. Production flow analysis assesses the operation sequence and the routing of the part through the machines in the plant. Using the data from operation sheets or route cards instead of part drawings, part families are formed. Part classification and coding is perhaps the most effective and widely used method. In this approach, parts are examined abstractly to identify generic features that are captured using an agreed-on classification and coding system. Though they are the most costly to implement, classification and coding systems are the most accurate.

Coding Systems. The two main coding systems in use today, *attribute-based* (polycodes) and *hierarchical-based* (monocodes), are shown in Figure 20.6.

In attribute coding, the simpler of the two, code symbols are independent of each other. Codes of fixed length span parts families, and each position in the code corresponds to the same variable. Because of this, each attribute to be coded must be represented by one digit, which can make the code quite long in some cases.

A hierarchical code structure is designed so that each digit in the sequence is dependent on the information carried in

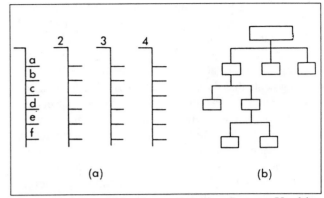

Figure 20.6 Two Main Types of Coding Systems Used in Group Technology: (a) Attribute Based and (b) Hierarchy Based.

the digit just preceding it. Generally, the first digit holds the most basic information, and each succeeding digit contains more specific information. This makes it possible to capture a great deal of information in a relatively short code.

20.4 Information Resource Management

Information is a major management issue in the manufacturing enterprise. In previous decades, information management was thought to be a problem that had to be solved by each organization for itself. Today, information is viewed as a management opportunity for the enterprise as a whole. Tremendous amounts of time and attention are devoted to understanding exactly what manufacturing information is and to developing new and better techniques and tools for managing it.

The concept of data management is at the heart of information resource management (IRM). Manufacturing productivity is believed to be linked to the notion of shared or common data, especially between engineering and manufacturing. One objective of IRM is to break down the walls traditionally existing between the two organizations; in effect, rendering them one organization through a common database.

Just as the design and manufacturing functions are interrelated, material movement and material management are interrelated. They may be accomplished by someone carrying a piece of metal or a box of bolts to a workbench where the two will be combined or by an MRP program generating a pick list and sending it electronically to an automated storage and retrieval system that, in turn, directs the physical movement of the selected parts to the appropriate workstation.

Computer-integrated manufacturing (CIM) is based on managing information resources using a common computerized database. The CIM wheel shown in Figure 20.7 illustrates the interrelationships among the manufacturing process—a single entity—and numerous segments and subsegments that do not and cannot operate without some impact from or on other segments. These impacts exist whether or not a computer is part of the manufacturing environment; the individual segments of a manufacturing organization are interdependent. Information resource management can be more effective using CIM; together, they help the various aspects of the manufacturing enterprise run smoothly as a whole.

20.5 Human Communication

The communication model, Figure 20.8, indicates a fairly simple process. The communicator and receiver are any participants in the organization at any level. The effectiveness of the communication can be evaluated by the degree to which the receiver obtains the message as intended by the sender. However, there are many barriers to communication. People may say one thing and appear to mean another because of nonverbal clues transmitted to the receiver. These might include body movement and facial expression, vocal characteristics, physical distance between sender and receiver, and time orientation. Other significant barriers include role perception in the company hierarchy between sender and receiver, selective listening, use of jargon, perceived credibility of the source, and "noise" and filtering based on personal agenda and other factors.

Organizational behavior is more complex than a compendium of individual behaviors. A manufacturing manager needs to understand how membership in a group affects individuals, how groups define individual roles, how norms are established and controlled, and how decisions are made by groups. Groups are either formal (individuals interacting based on their roles in the company hierarchy) or informal (individuals' interaction based on factors other than their organizational relationship, such as ethnicity, age, religious affiliation, or family). The organizational leader should come to know both informal and formal relationships and other factors such as relative status of individuals and groups. Group standards of conduct result in what is considered to be acceptable or unacceptable behavior. These norms and standards are frequently derived as a company or plant "culture." The culture may be closely linked to those values needed for company prosperity or it may conflict with enterprise goals. Organizational behavior must be viewed and understood as a result of compliance with the dominant cultural values. Good management

works toward developing a company culture that supports values and reinforces behaviors which serve the enterprise as well as the individual.

Graphic communication is an important complement to other techniques such as face-to-face oral communication or written communication via newsletters, etc. A person remembers about 10% of what is read, 20% of what is heard, 30% of what is seen, and 50% of what is seen and heard. Charts, slides, cartoons, transparencies, posters, and even videotaped presentations are significant means of conveying information to employees. On the manufacturing floor, visual cues for operators, trouble light (andon) boards, and graphically based job aids or operator instructions are all quick and effective means of communication.

20.6 Safety

Automated versus manual operations place a greater onus of responsibility on the engineer. In the manual operation, a worker initiates the change of state of equipment and/or process. In the automated mode, a controller of some sort (often remote from the areas of potential hazard) effects the change of state. Safe automated systems require more effort in the machine design phase and, increasingly, in the nature of communication interfaces within the factory. Also, automated systems require training for the operator which may include more sophisticated concepts than those for manual operations. To illustrate, Figure 20.9 depicts a fault-tree analysis of hazards created by robots.

Hazard awareness is a key element of safety training specifically and health and safety programs in general. Management must work with safety professionals and the work force to determine locations with a high accident potential as well as to identify severe hazards with a low occurrence likelihood. Hazard analysis precedes hazard awareness and should factor in both the probability of accident occurrence and the severity of injury or property damage. After identification of hazards has been completed, hazard control methods are implemented. Training in hazard awareness should be done after the hazards have been identified and analyzed and control measures undertaken.

The U.S. government's product-liability laws require that all products comply with appropriate standards and regulations and be free of safety-related defects. Lawsuits against manufacturers have mushroomed in recent years (in the U.S.). Companies have been forced to be proactive and defensive in anticipating liability claims based on injury or death caused by their products. Civil lawsuits for damages

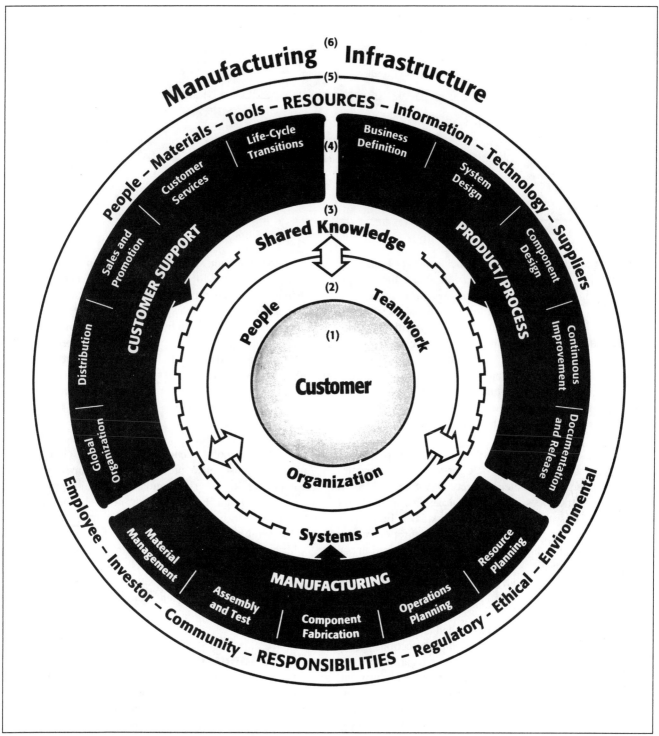

Figure 20.7 The CIM Wheel Emphasizes that CIM Encompasses the Total Manufacturing Enterprise. (*CASA/SME*)

predominate and are aimed at companies. However, criminal liability suits may be directed at corporations and, in a few cases, individuals. For the great majority of industrial managers, the probability of criminal prosection is remote. Before there can be such liability, the manager must be found guilty of knowingly carrying out illegal actions or being grossly negligent in his or her duties. Manufacturers have been under increasing pressure to exercise care and to document all activity in the design and manufacture of products.

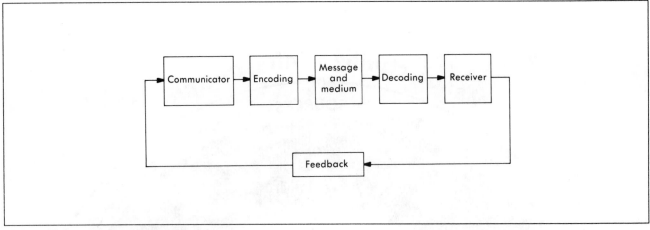

Figure 20.8 A Communication Model.

Lost-time accidents are those in which a worker loses time from the job, either immediately or at a later date. Typically, any injury requires some time lost for treatment, so an injury is usually considered lost-time if a worker fails to return the day after the accident. Safety records are commonly compared based on lost-time incidents. Non-lost-time accidents may be designated minor injuries and compensable injuries. Records of lost-time injuries also carry some determination of injury severity. Improvements in safety records would record the number of lost-time accidents in conjunction with a measure of severity.

Many states have "Right-to-Know" laws as part of their health and safety legislation. "Right-to-Know" mandates that employers make Material Safety Data Sheets (MSDS) available to employees, in a readily accessible manner, for those hazardous chemicals in their workplace. Employees cannot be discharged or discriminated against for exercising their rights, including the request for information on hazardous chemicals. Employees must be notified and given direction for locating Material Safety Data Sheets and the receipts of new or revised MSDS(s).

All hazardous materials must be labelled in accordance with federal MSDS. The U.S. Environmental Protection Agency has oversight of hazardous materials with regard to use and human exposure, labelling, containment, and disposal. Increasingly, the design-for-manufacture (DFM) process attempts to engineer hazardous materials out of the product and process. While many efforts such as DFM have been undertaken to reduce and/or eliminate these materials from the manufacturing environment, many are still present. Each plant is responsible for the management of hazardous materials and frequently a broad-based hazardous materials committee establishes procedures consistent with EPA regulation, state-level regulation, and company policy. Common procedures include control of exposure including operator protection such as mist collection systems, gloves, and guards; rules for storage, handling, and disposal; as well as contingency plans and training for accidental spills and exposure. Hazardous materials constitute a real danger to employees, attested to by recent research and codified by EPA regulations. As a result, hazardous materials control has become an onerous and expensive, yet necessary task for manufacturers. The clear trend in hazardous materials is to reduce their use wherever possible and engineer them out of products and processes.

Environmental concerns pervade the manufacturing workplace to the point where manufacturing engineers and managers must make protection of the environment and compliance with myriad regulations a major priority. The task is difficult but there are no real alternatives to proactive efforts to protect the environment, workers, and the general public. First, the manufacturer must determine which codes must be met and the nature of the process to issue permits and submit reports. Areas of major concern include air and water pollution, hazardous wastes, existing conditions, and noise amendments.

20.7 Manufacturing Management

Project management begins with a good understanding of the product and the environment in which the project will be realized. Knowledge of the technologies involved and of the financial and contractual matters is also essential. Strong human relations and communications skills are also critical. A project refers to all activities associated with the achievement of a set of specific objectives considered worthy of financial support. It has a finite life, and its successful management requires that technical objectives be met, project timelines are adhered to, and that the budget not be exceeded; in short, completed per specifications, on

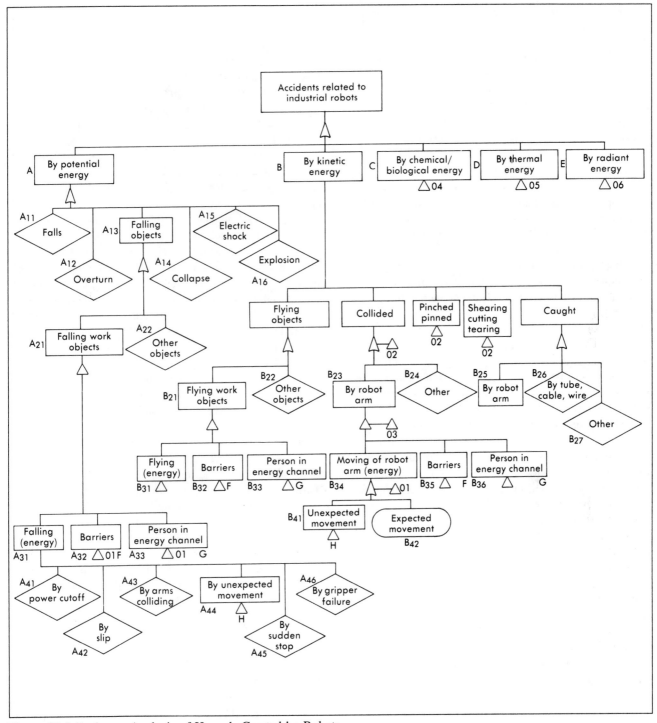

Figure 20.9 Fault-tree Analysis of Hazards Created by Robots.

time, and under budget. Milestones to check progress are normally executed at logical points in time and some form of summative evaluation at the end is usually done. Many project control tools are available to the project manager, such as Program Evaluation and Review Technique (PERT), Critical Path Method (CPM), and Gantt charts.

Problem-solving allows a manufacturer to get more outputs using the same or fewer inputs. Improving productivity and quality and lowering costs can be done by harnessing the thinking skills of managers and workers. Improving the quality and precision of problem-solving in an organization releases a productive force far greater than that of equipment, tools, software, and more worker effort. The general steps for problem-solving are problem recognition, problem definition or specification, developing possible causes, testing for most-probable causes, and verification that the problem is solved. Many techniques are used, almost always in a group setting. Common problem-solving tools include Ishikawa or fishbone diagrams, Pareto charts, cause-and-effect analysis, brainstorming, force-field analysis, and statistical tools such as histograms and control charts.

Leadership may be viewed from the "big" picture, as an individual who has brilliant ideas and the capacity to inspire others, or on a more pragmatic level as a person with the interpersonal influence used in face-to-face dealings with subordinates, superiors, peers, and others who may have an effect on the organization. Measures of leadership must assess both getting the job done and the effects of getting the job done on others and the organization. The managerial grid shown in Figure 20.10 illustrates some key leadership elements for manufacturing managers.

Planning has been considered a key responsibility for management for most of the last century but has taken on more importance in the more recent period of rapid change. Manufacturers today must be prepared for and anticipate surprises and crises, promote flexibility and adapt to unanticipated changes, identify new business opportunities, identify key problem areas, foster motivation, enhance the generation of new ideas, communicate top management's expectations down the line, foster management control, promote organizational learning, communicate line managers' concerns to top management, integrate diverse functions and operations, and enhance innovation. In a given organizational context, generic planning and control systems consist of several key elements: a plan or desired state, actual performance, and controls that compare the plan with actual performance and suggest changes. Such a "plan, do, review" sequence is typical in many manufacturing organizations.

Some of the major concepts of management of the organization that have emerged include centralization-decentralization, line-and-staff, and span of management (also known as span of control). Centralization-decentralization has to do with where real authority resides. That is, if authority is not delegated but rather resides in one person, then the organization can be regarded as centralized. Decentralization is the extent to which authority is delegated. Measurements of centralization-decentralization are qualitative rather than quantitative and can be made by assessing decision-making. The more decisions made at the lower levels, the more decentralized the organization; the more important the decisions made at lower ranks, the more

decentralized; the broader the types of decisions made at lower levels, the more decentralized.

Line-and-staff refers to relationships within an organization. Line relationships are those established by the flow of authority. Staff relationships are advisory. Staff departments in a manufacturing plant are those units which contribute advice such as financial, legal, or engineering advice. Line departments are those performing activities critical to the smooth operation of the organization. In a production facility common line positions are plant manager, area managers, superintendents, and first-line supervision.

A budget is a financial plan for an organizational unit. A budget should permit planning, coordinating, and controlling the flow of capital for a given unit in conjunction with divisional or company budgets. The budgeting process in manufacturing involves the recovery of costs such as indirect labor, overhead, and general/administrative costs. These are costs of plant operations which support direct labor. The recovery is accomplished by charging an hourly rate for each direct labor hour worked.

Management control is the process by which managers ensure that resources are obtained and used effectively and efficiently in achieving the organization's objectives. The function of management control is to facilitate reaching organizational goals by implementing strategies identified in the planning process. The control process must be accomplishable through the use of technology and people. Three important aspects are pragmatism, results, and people. Controlling is maintaining behavior within preselected parameters. Manufacturing converts the resources and adds value to them. The controlling subsystem ensures that the product meets quality standards and schedules. A management control system allows an organization to record, measure, and control variability in the business process against predefined goals and objectives.

20.8 Organizational/Industrial Psychology

Teamwork. When personal and enterprise goals are congruent, motivation is unquestionably higher. Employee participation in decision-making which affects their future imbues a sense of belonging and of empowerment. A common and successful vehicle for such meaningful participation is the work team. The team may be temporary or semi-permanent. Team members support one another and can create a synergy in problem-solving resulting in superior performance. An individual's need for social interaction, for approval and esteem from others, for personal

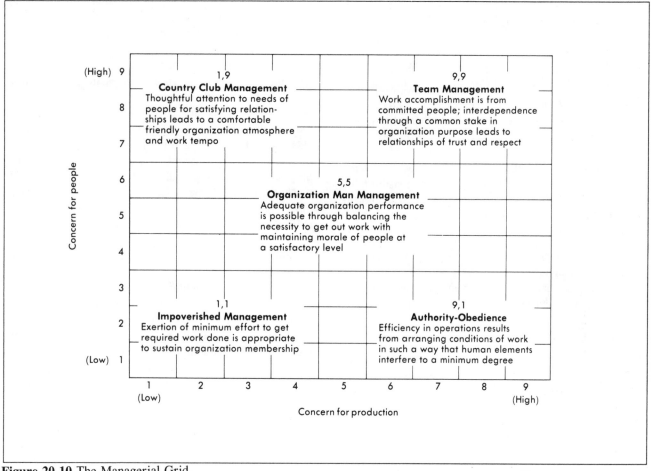

Figure 20.10 The Managerial Grid.

achievement, and for self-realization can all be met in a well-managed, team-oriented environment.

W. Edwards Deming: A Theory for Management

Transformation Through Application of the Fourteen Points

1. Create constancy of purpose toward improvement of product and service, with the aim to become competitive and to stay in business, and to provide jobs.

2. Adopt a new philosophy. We are in a new economic age. Western management must awaken to the challenge, must learn their responsibilities, and take on leadership for change.

3. Cease dependence on inspection to achieve quality. Eliminate the need for inspection on a mass basis by building quality into the product in the first place.

4. End the practice of awarding business on the basis of price tag. Instead, minimize total cost. Move toward a single supplier for any one item, on a long-term relationship of loyalty and trust.

5. Improve constantly and forever the system of production and service, to improve quality and productivity, and thus constantly decrease costs.

6. Institute training on the job.

7. Institute leadership (see point 12). The aim of leadership should be to help people and machines and gadgets to do a better job. Leadership of management, as well as leadership of production workers, is in need of overhaul.

8. Drive out fear so that everyone may work effectively for the company.

9. Break down barriers between departments. People in research, design, sales, and production must work as a team, to foresee problems of production and in use that may be encountered with the product or service.

195

10. Eliminate slogans, exhortations, and targets for the work force asking for zero defects and new levels of productivity.

11. a. Eliminate work standards (quotas) on the factory floor. Substitute leadership.

    b. Eliminate management by objective. Eliminate management by numbers, numerical goals. Substitute leadership.

12. a. Remove barriers that rob the hourly worker of his right to pride of workmanship. The responsibility of supervisors must be changed from sheer number to quality.

    b. Remove barriers that rob people in management and in engineering of their right to pride of workmanship. This means, *inter alia*, abolishment of the annual or merit rating system and of management by objective, management by the numbers.

13. Institute a vigorous program of education and self-improvement.

14. Put everybody in the company to work to accomplish the transformation. The transformation is everybody's job.

Participative management techniques are based on power sharing among managers who, in a traditional command-and-control organization, are the sole executors of power. This form of organization is based on the premise that group participative problem-solving and decision-making contribute to higher quality problem solutions while increasing the level of worker commitment. Specific techniques include quality circles, kaizen groups, employee involvement groups, cross-functional problem-solving groups, etc. Organizations effectively utilizing worker participation in problem-solving have means of passing upwards problems requiring management decisions. Important characteristics of participative management systems include unwavering commitment to the process by the highest levels of management; effective and constant communication up, down, and sideways throughout the organization; trust between management and the work force; and true empowerment of employees to implement their decisions. Participative management is consistent with the view that workers are important assets of an organization and that most people want to work and get satisfaction from their work by doing a good job. This is in contrast to the description of traditional management techniques which assume that labor is an interchangeable commodity like any other purchasable resource and that most people find work distasteful and will avoid it if possible. Participative management can harness the energy of individuals by allowing their personal achievement to advance simultaneously with the goals of the organization.

Goal-attainment or goal-setting theories of motivation suggest that individuals with specific goals are better performers than those with no goals. Within this theory, the behaviorist (or operant condition) approach states that individuals want rewards for attaining goals and that managers should see that such rewards are provided for in the organization. A similar view says that when an individual is involved in setting the goals, more effort will be made to reach them. Thus, participation in the creation of the objectives is itself a motivator. Management-by-objective (MBO) is one example of goal-setting theory.

Incremental, continuous improvement of manufacturing processes is one of the keys to the Japanese manufacturing miracle. Japanese manufacturing companies are likely to spend two-thirds of their R&D budgets on process improvement and one-third on product improvement. For U.S. companies, just the reverse is true. Improving quality, reducing costs, reducing the concept-to-market time, and turning inventories more quickly are all areas of potential improvement. U.S. companies tend to view a manufacturing process as static after ramping up to production levels, while the Japanese or "continuous improvement" model sees the process as dynamic with the quest for improvement never ending.

20.9 Supervision

Supervision is the motivation and guidance of subordinates toward goals established by the enterprise. Historically, this role has been the first level of management with direct oversight over factory workers. While the traditional "command-and-control" supervisor still abounds, manufacturing organizations are moving to supplant this role by hourly team leaders and supervisors who act as coaches and facilitators rather than autocrats.

Supervision of technical employees is distinct from that of production employees. With technical employees such as engineering technicians or even manufacturing engineers, the supervision is much less direct; these employees exercise much more autonomy. They need less direction and more support and liaison work. However, the supervisor is still responsible for the work being done and has additional work not required of first-line supervision such as budgets and personnel evaluation.

Organizational leadership styles may be viewed through the managerial grid of Figure 20.10 in which the concern for production is juxtaposed with the concern for people.

Basic people skills make up the vertical axis of the managerial grid. To be an effective coach or motivator, a manager must be effective in relating to individuals and groups. Skills such as one-to-one communication, group facilitation, providing support and feedback for performance, and listening skills are among the most important people skills.

Reporting hierarchy typically refers to the line relationships referred to in the Management section of the chapter. Reporting hierarchies are usually pyramid-shaped and define supervisor-subordinate roles as well as the formal up-and-down flow of communications in an organization.

20.10 Facilities Planning

Plant Layout. Plant layout may be defined as the planning and integration of the paths of the component parts of a product to obtain the most effective and economical interrelationship between employees, equipment, and the movement of materials from receiving, through fabrication, to the shipment of the finished product.

Facility layout is critical to the productivity of a plant. If facility layout is *not* optimal, it affects the nonoperation time for the part, the level of manpower required to move the part, and the capital investment in material handling equipment. First, it increases the nonoperation time (T_{no}), which is a component of manufacturing lead time. This, in turn, lengthens manufacturing lead time, increases work-in-process inventory, and increases the capital investment in work-in-process. Second, the increase in nonvalue-adding material handling adds man hours to the cost of the

product without adding any value to the product. Finally, the capital investment in material handling equipment is increased commensurate with the greater material handling activity without generating any more income from the sale of the product.

Facility layouts are of the process, product-process (cellular), or fixed (station) types. The facility layout may be designed according to function or process. The process-oriented plant layout is most common in manufacturing today. This arrangement groups together all similar functions such as milling, turning, grinding, etc., resulting in an arrangement that requires less capital, achieves higher machine utilization, and is easier to automate. A typical process facility layout is shown in Figure 20.11.

In the product-process arrangement, one product family is produced in a given cell using group technology. This arrangement, as shown in Figure 20.12, produces greater volumes of the part family with shorter manufacturing lead times; the batch of parts is not waiting to be moved to the next process (T_{no}) and the next setup (T^{su}). The equipment for the product family is arranged either linearly or radially according to the sequence of processes. The process cell has the least material handling and work-in-process inventory. Continuous operation of the product process is easiest to automate with robots and other material handling equipment.

The fixed (station) type of production, Figure 20.13, has a fixed or stationary product with the manufacturing and assembly going on around it. This layout is typical of large, low-volume products such as machine tools and aircraft. This layout is flexible, adaptable, uses the smallest amount of skilled labor, and requires a minimal amount of product planning.

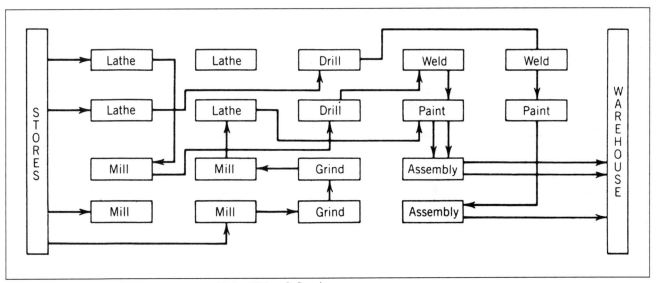

Figure 20.11 Process Facility Layout. (*John Wiley & Sons*)

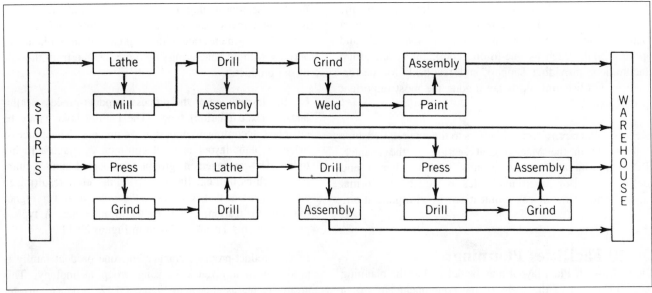

Figure 20.12 Product-process (Cellular) Layout. (*John Wiley & Sons*)

In many cases, manufacturing processes and layouts are designed with the material handling function completed as a means to accomplish the production. This approach typically results in inefficient material handling and production systems. Material handling should be an integral part of the production processes and facility planning effort.

Of concern is what must be done to position the new machine on the floor to minimize production costs and ensure quality levels and how the process engineer can maximize the time spent in adding value to the part. A number of major criteria should shape the equipment

acquisition and installation process as it pertains to plant layout. The position of the machine on the plant floor should facilitate the manufacturing process and maximize machine utilization. Material handling should be minimized, including work handling to change positions for machining. Some flexibility of arrangement and operation should be maintained. Work-in-process should be minimized through a high turnover rate. Equipment investment should be minimized through effective use of floor space. Direct and indirect labor should be minimized and a safe and convenient workspace should be created. A variety of production components is typically affected by plant layout.

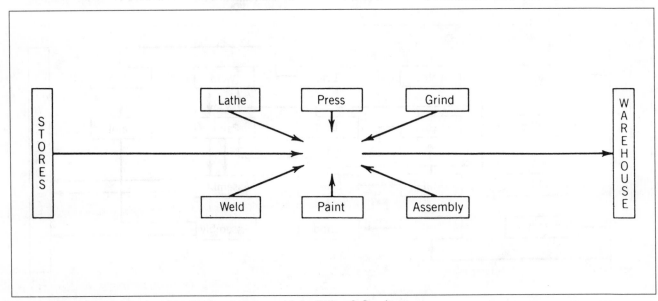

Figure 20.13 Fixed (Station) Production Layout. (*John Wiley & Sons*)

198

Location and type of external transportation must be considered:

- Are parts arriving by truck or rail?

- What is the form of packaging of incoming stock?

- Are there intermediate steps between external transportation and the operation being planned such as unpacking, inspection, or storage?

The nature of the production activities must be established as well as their location. The position of service and auxiliary activities must also be considered. For example, the location of extra tools and maintenance stations should be considered. The location of the new machine and gaging or other inspection stations must be carefully planned. Optimal location should be planned for buffer storage or shipping on the output side of the process. Basic guidelines for effective plant layout include:

- A planned materials flow pattern,

- Adequate, straight aisles,

- Minimal backtracking,

- Locating related operations close together,

- Predictable production time,

- Minimal work-in-process,

- Some built-in flexibility,

- Maximum ratio of processing time to overall time,

- Minimum distances for material handling,

- Optimal quality practices,

- Minimal manual handling,

- Materials in unit loads,

- Receipt of materials by production employees,

- Smooth and adequate materials flow,

- Good housekeeping,

- Planned task responsibilities that balance employee work loads, and

- Access to maintenance.

20.11 Maintenance

There are fundamentally two approaches to the maintenance of production equipment: complete repairs when something breaks down (corrective maintenance) or scheduled preventive maintenance (PM). PM will not prevent all breakdowns, but can dramatically reduce the amount of production time lost waiting for equipment to be repaired. The main activities of PM are lubrication, parts replacement, machine adjustments, and overall inspections. PM may also include regularly scheduled input by machine operators to describe potential problems. With PM:

- Machines will last longer,

- Maintenance time and cost will be cut,

- Severity and frequency of breakdowns will be reduced,

- Safety levels will rise,

- Product quality will be maintained, and

- Production costs will be cut by increasing asset utilization time and decreasing time lost by idle operators.

A more sophisticated form of preventive maintenance is predictive maintenance which involves the use of various types of sensors to predict breakdowns. Monitoring vibration signatures is a common example of predictive maintenance. A baseline vibration signature is recorded when the machine is set up so that variation from this signature signals malfunction. The widespread use of sophisticated sensors, programmable controllers, and other computers has made predictive maintenance a bit easier to execute.

A well designed preventive maintenance program includes:

- Adequate record keeping,

- A regular schedule of inspection to determine optimal intervals between inspections,

- Use of checklists,

- Well qualified tradespeople as inspectors,

- Appropriate budget allocation,

- Administrative procedures to ensure compliance and followup, and

- Input by machine operators who know the machine and/or process the best.

Preventive maintenance can begin at the first stage of machine design when features that reduce the need for maintenance can be integrated. Machine design can include ease of access for maintenance people to execute PM. In the process of machine specification and acquisition, the concept of preventive maintenance should be a prominent factor. Proper location of lubrication points or reservoirs, design of guards, access to motors and fluid power components, ease of cleaning, proper chip disposal, design of lockouts, ease of visual inspection of critical areas, and accessibility of adjustment points are just some of the many elements enabling successful PM which should be included in a machine design. Typical preventive maintenance guidelines include recommended lubricants, checks and check frequency, and PM procedures. A widely used predictive maintenance procedure is that of a baseline vibration signature using a fast fourier transform (FFT) device.

Statistical Process Control (SPC) monitors piece part characteristics such as diameters, hole location, etc. SPC charts such as X-bar and R charts inform the operator if the process is "in control"; that is, the variation is "normal" or due to "assignable causes." If a process (which has been found to be capable) is in control, then "normal" process variation yields parts within the allowed tolerance band. However, if a process (which has been found to be capable) is out of control, it is because of assignable variation and bad parts are or will be produced. Assignable causes may be tool wear, machine vibration, out-of-spec material, changes in the environment such as temperature, operator error, etc. Good preventive and predictive maintenance procedures should reduce the likelihood that assignable causes of excessive process variation are due to poor equipment condition. Some SPC auditing systems require documentation of such procedures to be certified as a parts supplier.

20.12 Quality

TQM Concepts. Total Quality Management (TQM) is a philosophy derived from the work of individuals such as Shewhart, Deming, Juran, and Feigenbaum. TQM is a philosophy which states everyone shares the responsibility for an organization's quality of performance and output. Typical goals of TQM include: first-run capability or "doing it right the first time," high levels of employee involvement, continuous improvement of quality and productivity, and total maintenance. These goals apply not just in product but also in supplier and customer relations.

Production design quality assurance programs should accommodate information from customers as well as company policy and should be initiated with a detailed design and validation plan. The cross-functional product design group will document appropriate specifications on the product drawings. The most widely used means of communicating these specifications is the geometric dimensioning and tolerancing system (ANSI Y14.5M). Design validation will follow the creation of initial specifications and use tools such as calculation and design failure mode and effects analysis (DFMEA), fault-tree analysis, or other tools. A failure mode identified through DFMEA may be eliminated or modified through a design change, or a manufacturing or quality procedure may be instituted to minimize risk of failure. Tests of prototypes or early production models should be made to further anticipate problems prior to full production. Design reviews may call for changes to resolve problems flushed out by the DFMEA or testing.

Manufacturing process quality systems are implemented to meet design specifications through quality control. Using a technique similar to DFMEA, a process FMEA may be done to anticipate process-related quality problems. Many manufacturers classify part features into levels of importance for product safety, function, and appearance. Quality plans then emanate from the level of importance of a dimension or other characteristic. Manufacturing control plans also include some type of "runoff" to verify machine process capability. Gage capability is also an important part of quality from the manufacturing perspective. Gage R and R determines accuracy and capability of a gage. Inspection plans will also be determined along with sampling rates.

Metrology and measurement in a production environment are executed primarily using gages designed for a specific use, usually a single dimension. General purpose measuring tools may be used but are generally considered too slow and prone to operator error to measure high-volume production. Gages are used for either attribute (go/no-go) or variable (continuous value) measurement.

The precision and accuracy of a gaging/metrology system is as important as the manufacturing process itself in ensuring part quality. Permissible gage tolerance is frequently expressed as a percentage of the dimension's tolerance. The smaller the tolerance used by the gage, the higher the level of precision and, consequently, cost. As a rule, the gage should take up no more than 10% of the dimension being gaged. Standard gagemakers' tolerances have been established for certain types of gages such as limit-type cylindrical plug and ring gages. Four classes—XX, X, Y, and Z—are used, with XX as the most accurate.

Measurement methods are infinite in their variety. Major categories include mechanical indicating gages, pneumatic and pneumatic/electronic gages, electronic gages, gage blocks, fixed functional gages such as plug and ring gages, spline and gear checkers, comparators, lasers, profilometers for surface measurement, coordinate measuring machines (CMM), machine vision systems, and so on.

Automated gaging may assure greater productivity, lower costs, and higher quality. The variability of operators can also be eliminated. Practically any dimension can be checked. Automated measurement and gaging systems may easily be connected to computers and statistical process control, or other software packages may be used. Automated measurement systems may be used for pre-process, in-process, post-process, and/or final inspection.

Practice Problems

20.1 Three employees at a management level beneath the policy-making level in a high-tech organization—a sales director, the head of research and development, and the manufacturing manager—engage in a conspiracy to steal product secrets from a competitor. Budgets are quietly manipulated and persons are hired as spies to ferret out the innovative design/production/distribution methods that give the competition such an edge. Listening devices may also be employed. Such a deed is not normal business and goes well beyond the bounds of ethical behavior. If caught and prosecuted, who is considered ultimately legally responsible?

20.2 Conduct a simple value analysis by estimating the relative percentages of the functional and esteem values, and waste, for a common metal paper clip.

20.3 Calculate the EOQ for a product that has an annual usage of 100,000, setup and order costs of $50, and an interest and storage cost of 20%.

20.4 A person retains how much of what is heard?

   a. 10%
   b. 20%
   c. 30%
   d. 50%

20.5 Employee "Right-to-Know" laws focus on

   a. Dangerous machinery
   b. Affirmative Action plans
   c. Hazardous chemicals
   d. Company policies and procedures

20.6 The management concept in which many of the decisions affecting the operation of the plant are made at the lower levels is called

   a. Centralization
   b. Line-and-staff
   c. Span-of-control
   d. Decentralization

20.7 Quality circles, kaizen groups, and cross-functional problem-solving groups best describe what type of management technique?

   a. Command-and-control
   b. Participative
   c. Top-down
   d. Traditional

20.8 What differentiates predictive maintenance from preventive maintenance?

COMPUTER APPLICATIONS/ AUTOMATION

By the mid-1980s four criteria for automating manufacturing had become solidified: to improve (1) product quality, (2) production efficiency, (3) quality of work life, and (4) product development time. Production efficiency applies to both fixed and flexible automation and includes but does not focus on labor replacement and reduced costs. The improvement of quality is important because of the cost of both the raw materials (approximately 40% of the cost of the final product) and the value added to the raw materials in both operator and machine time. In most manufacturing situations today, replacement of labor is not an economically viable rationale for automation because direct labor is a small component (5% to 15%) of the final cost of the product. However, improving production efficiency may help current production personnel improve the direct labor-to-product ratio. The improvement of the quality of work life minimizes most of labor's concerns pertaining to poor working conditions, high workmen's compensation claims, high absenteeism, high employee turnover, and low product quality. The improvement in product development time enhances competitiveness by reducing the time from product concept to its launch into the market. This criterion cuts all engineering times but does not alter production itself.

Computer applications in manufacturing consist of using the computer to process data and information. Computer simulation is an efficient tool for planning manufacturing systems. Material requirement planning (MRP) and manufacturing resource planning systems (MRP II) manage manufacturing systems by processing both data and information to monitor and control production. Specific problem-solving and information processing are completed with spreadsheets/databases. Computer-aided design (CAD) and computer-aided manufacturing (CAM) are fundamental tools in cutting-edge design, engineering, and manufacturing.

Material requirement planning, manufacturing resource planning, master productions scheduling (MPS), and quality control comprise the manufacturing management dimension. For all but the smallest companies, MRP (including MRP II) is done by computer.

Computer-aided design and computer-aided engineering (CAE) obviously are part of the beginning of a product life cycle. Computer-aided process planning (CAPP), cost estimating (CACE), as well as programmable and flexible production equipment with computer control systems (e.g., computer numerical control (CNC), programmable logic control (PLC), and robotics) relate to the physical manufacture of products. The technologies that tie all of these together include networking and integration of computer systems through hardware and software.

Integration of computers in manufacturing equipment generally consists of programmable or flexible automated equipment, but hard automation increasingly is incorporating programmable automation as components, in the form of CNC machine tools and PLC as computer-based equipment makes gains in reliability.

21.1 Simulation

Simulation is a process of experimenting with a model of a real system. The experimentation involves changing the structure, environment, or underlying assumptions of the model. Functions lending themselves easily to simulation are specific processes, product performance, and discrete event performance. Simulation software does not optimize a process or system, but it can model (estimate) performance in a reduced period of time. A realistic simulation program requires specific information, structured and specified in the form accepted by the software. This characteristic of simulation software helps users to be thorough in the planning process. In some cases, simulation may be helpful simply because it forces users to identify relevant data needed to execute the simulation.

Simulation related to manufacturing includes, but is not limited to, computer/network performance, CNC programs, machined surface finish, product performance in various environments, kinematics of workcells (including robot programs), manufacturing systems, and continuous processes. A desirable feature of these simulation activities is that the manufacturing equipment, cells, or system are not taken out of production for program verification and debugging. Engineers and operators can have a high degree of certainty about the performance of a machine, cell, program, or system when the simulation is implemented. Simulation also provides the capability to execute "what if?" scenarios as part of the improvement of production efficiency.

Simulation of product performance can be a factor in reducing the development time for products while still achieving the desired performance level. Simulation of product performance reduces the number of prototypes to be built and tests to be conducted by accomplishing many of these tasks in software as opposed to hardware. This type of automation satisfies the criterion for automation of shortening the product development time.

Simulation of CNC machine tool performance can verify tool paths and tooling for the program without taking the actual machine tool out of production. Other software programs simulate tooling, tool paths, the CNC program, and workpiece material and generate surface finish performance.

Static simulation, as the term implies, is performed without consideration of time. Spread sheets are usually appropriate for this type of simulation. Dynamic simulation is time-based and requires that system performance with respect to time be taken into account. Kinematic simulation of robotic programs and manufacturing cells and discrete event manufacturing systems are typical dynamic simulations.

Simulation of discrete event manufacturing systems enables the evaluation of a manufacturing system for production routing, scheduling, buffers and queues, resources, and layout. Manufacturing lead time, throughput, work-in-process, and bottlenecks are examples of the application of this type of simulation. The modeling of manufacturing elements includes parts to be produced, resources (equipment and personnel), and production data used to produce the parts. Animation, another attribute of simulation, is valuable for illustrating specific issues for the simulation team and for presenting issues to management.

Deterministic simulation is based on statistical means for the respective events in the system. It presumes the actual values do not vary from the entered values. Stochastic simulation includes the capability to generate models based on statistical distributions or random numbers appropriate to each discrete event within limits of the specified statistical parameters and to provide output data for the system. Stochastic models should be executed a number of times (generally without animation) to determine the output data related to a stochastic input model. The stochastic evaluation of a dynamic system generally yields more accurate results.

Continuous process simulation focuses on production of continuous processes such as those in the chemical industry, heat exchangers, cooling towers, and pumps. Since the processes are continuous, no discrete events occur over time. Startup, steady-state operation, disruptions to the system in the form of component malfunctions, and shutdown are all events to be simulated.

21.2 Material Requirement Planning

Material requirement planning has evolved over the years into a time-phased system of controlling inventory for manufacturing. It is based on bills of material (BOM), for production specified by the master production schedule, and the current inventory with outputs of purchase orders and shop floor release orders for production. MRP has evolved into a fully integrated manufacturing resource planning system, MRP II. MRP II includes all of MRP but also integrates capacity requirements planning (CRP), production planning, and production activity control. The issue of MRP versus MRP II is significant in helping manufacturing managers determine the type of system or software they need and the integration of the various manufacturing and management areas into the system. MRP and MRP II are designed for batch manufacturing a variety of products in a *push* system. MRP and MRP II are executed on both mainframe and microcomputers. There are various levels of MRP and MRP II with Class A compliance (the highest level) consisting of 95% accuracy in inventory and 98% accuracy for BOM. A company's use of MRP or MRP II will not guarantee improved on-time deliveries, improved productivity, reduced costs, or reduced inventories, but it is a valuable component of a successful business strategy to accomplish these goals.

Companies with continuous production of a few products with simple bills of material probably do not need MRP. In very small manufacturing companies, MRP functions can be executed with pencil and paper, but in larger companies computer-aided management (i.e., MRP or MRP II) is required.

The MPS drives the MRP system based on customer orders or sales forecasts. The BOM represents the required parts and materials used in manufacturing a product to the MRP system and the inventory control data reports the existing inventory to the MRP system. Since MRP is a time-phased system, both manufacturing and ordering lead time are important issues. Lead time is a factor in scheduling (production queue), ordering (parts and materials), and in manufacturing (setup, operation, and non-operation time) inasmuch as the MRP system issues purchase orders and production orders.

MRP II integrates sales forecasts with BOM, inventory, CRP, and shop-floor control. Three issues in MRP II are (1) dependent versus independent demand, (2) manufacturing and ordering lead times, and (3) common use items. Each of these affects the operation of MRP II. If the variables are stable (or in control, just as production processes may be in control) the MRP II system can perform well. If the variables are unstable or out of control, MRP II will be problematic.

Dependent demand occurs when a component of one product is part of one or more other products. Parts and raw materials common to two or more products, then, are examples of dependent demand. Independent demand refers to parts or products that are not used in any other products. Spare parts and end products are examples of items that are independent of other products or parts. Independent demand items require forecasting or just-in-time scheduling, while dependent demand is calculated from known delivery schedules.

Lead time involves manufacturing lead time (MLT) and ordering lead time (OLT). Manufacturing lead time is used with both batch and mass production. Since setup time is more common to batch production, manufacturing lead time is more complex for batch production. Once the setups are made in mass production, lead times stay very consistent.

Batch production has an MLT that is a series of summations. The series of setup times for each operation plus the production operation time for each part at each operation and the non-operation time of material handling inspection, storage, and wait for first part checkoff are sequential, not simultaneous.

$$T_{mlt} = \sum_{i=1 \text{ to } j}^{n=m} [T_{sui} + (Q * T_{oi}) + T_{noi}]$$

where:
T_{mlt} = Manufacturing lead time in hours
$\sum_{i=1 \text{ to } j}$ = Summation of the times per setup per operation/workcenter
n = Number of operations in the production of the product
T_{sui} = Setup time (hours) required for individual (i) work centers
Q = Number of parts per batch
T_{oi} = Production time (hours) per part
T_{noi} = Non-operation time per workcenter (material handling, inspection, and storage)

The setup time per workcenter and the non-operation time per workcenter are not affected by batch size. The non-operation time often may be generalized to 16 hours to move the batch between departments, wait for a setup person, and wait for first part checkoff; 8 hours within departments; and 4 hours at the same workcenter but with a different setup. The individual setup and non-operation times are affected by the time required and by the number of the work centers. The production time is a product of the quantity of parts, the production time, and the number of workcenters. Both setup and non-operation times should be minimized by automation in general and flexible manufacturing cells and flexible manufacturing systems in specific.

For mass production, the lead time from start to end is the operation time of each workcenter plus the non-operation (material handling, inspecting, and storage buffers) time between workcenters.

$$T_{mlt-mp} = \sum_{i=1 \text{ to } j}^{n=m} [(T_{pi}) + T_{noi}]$$

where:
T_{mlt} = Manufacturing lead time in hours
$\sum_{i=1 \text{ to } n}^{n=m}$ = Summation of the times per setup per operation/workcenter
n = Number of operations in the production of the product
T_{sui} = Setup time (hours) required for individual (i) work centers
Q = Number of parts per batch
T_{oi} = Production time (hours) per part
T_{noi} = Non-operation time per workcenter

Ordering lead time is the time between the ordering point and the time the material is in inventory. OLT is a summation of the specific time to signal purchasing that the material is required, to release the purchase order, and the actual receipt of the material. An individual company's response time from order receipt to shipment is a variable that changes from supplier to supplier. Suppliers to manufacturing companies that have achieved a "quality" rating by the manufacturer often use computers for order entry and generating order pick instructions, packing lists, shipping labels, and invoices. Also, they generally specify that the products be shipped by an express company within the same working day.

When lead times are quantified, the largest individual contributors may be systematically reduced to improve manufacturing lead time. Quick setup/modular tooling may be developed to reduce the setup time and improve the quality of the product. When the process is known to be under control, inspection samples and the T_{no} may be reduced. It may even be that the operator can perform the setup and reduce T_{no} even further by not requiring a skilled tradesman of a different union grade to do the work. Cellular manufacturing generally will reduce the T_{no} to approximately zero.

Common use items are raw materials that are used by a variety of products. The MRP II system should integrate the common use item materials from the master production schedule, bills of material, and the inventory requirements of the master production schedule before issuing the purchase order or the time-phased purchase notice. This feature can enhance economic order quantity purchasing and minimize material shortages.

Dependent and independent demand parts affect an MRP II system in different ways. Materials for dependent demand are an additional feature that should be included in the calculations for determining inventory and purchase order or time-phased delivery quantities. The MRP lead time obviously is affected by both the MLT and the OLT. The OLT is also affected by the supplier's response time. The common-use items impact the size of the quantities ordered with both purchase orders and time-phased deliveries.

21.3 **Just-in-time**

Just-in-time is a business strategy that strives to match production to demand (orders). This order-production relationship is targeted toward reducing or eliminating waste commonly found in the form of rejected parts, lost production time due to equipment failures (unscheduled maintenance), non-value-adding activities (excessive material handling), and long setup times.

Software to enhance this form of production may be modified from either MRP or MRP II to accomplish JIT.

Depending on the flexibility of the particular MRP/MRP II software modified, it may be satisfactory. Software that is written primarily for JIT applications may address JIT issues in a more robust manner. The issues of manufacturing lead times, ordering lead times, and especially control of the manufacturing process and equipment maintenance and reliability are significant to being able to produce in a JIT mode. Also, preventive maintenance to achieve reliable equipment that can produce products within parameters established by quality process control is essential to JIT.

21.4 **Spreadsheets**

Engineering cost-estimations, iterative manufacturing lead times for batch productions, and static simulations are easily set up and solved with spreadsheet software. Spreadsheet software can preform database, financial, logical, mathematical, matrix, spreadsheet operative, statistical, string, and temporal functions. Macros are also available for frequently used patterns of keystrokes. Most spreadsheets are menu-driven and new features can be learned as the user needs them.

A spreadsheet example with values is depicted in Figure 21.1 and formulas in Figure 21.2.

21.5 **Database**

Database systems are superior to file processing systems when the files for application software contain records stored in individual files. Since the records are accessed by application software, there are often multiple files with the same records for the different application programs. Databases can reduce redundancy and inconsistency in data, difficulty in accessing data, isolation of data in little-used files, supervision of data, security of data, and integrity of data. Databases are a representation or model of reality. The development of databases is an iterative process, and with each iteration, the database should be a better model of reality. A good database is organized so that the only questions it cannot answer are questions that are never asked.

Databases must be managed by programs so that most people have access to the desired data. These management systems operate at the *physical level*, the *conceptual level*, and the *view level*. The physical level defines how the data is stored. Most users of databases do not operate at this level. The conceptual level defines the data that is stored in the database and the relationships of the data. This is the first level of abstraction used to determine what data should be stored in a database. The view level is the highest level

	Oper.	#	Machine	Date of Setup	Time hours	Setup $/Hour	Cost of Machine Downtime $ / Hour	Total Cost of Setup, $
2	MLT COSTS							
3	Current Date:			24-Aug-93			Cost of	
4							Machine	Total
5				Date of	Time	Setup	Downtime	Cost of
6	Oper.	#	Machine	Setup	hours	$/Hour	$ / Hour	Setup, $
8	Setup	4	Mill #2-02	24-Mar-93	3.30	$50.00	$75.00	$412.50
9				23-Apr-93	3.60	$50.00	$75.00	$450.00
10				22-May-93	2.90	$50.00	$75.00	$362.50
11				08-Jun-93	4.00	$50.00	$75.00	$500.00
12				01-Jul-93	3.70	$50.00	$75.00	$462.50
13				15-Aug-93	3.50	$50.00	$75.00	$437.50
15				Mean	3.50		Mean	$437.50
16				Std Dev.	0.34			
17				Range	1.10			

Figure 21-1. Spreadsheet showing MLT Data

	Oper.	#	Machine	Date of Setup	Time hours	Setup $/Hour	Cost of Machine Downtime $ / Hour	Total Cost of Setup, $
2	MLT COSTS							
3	Current Date:			@DATE(93,8,24)			Cost of	
4							Machine	Total
5				Date of	Time	Setup	Downtime	Cost of
6	Oper.	#	Machine	Setup	hours	$/Hour	$ / Hour	Setup, $
8	Setup	4	Mill #2-02	@DATE(93,3,24)	3.30	$50.00	$75.00	+((K8*M8)+(K8*O8))
9				@DATE(93,4,23)	3.60	$50.00	$75.00	+((K9*M9)+(K9*O9))
10				@DATE(93,5,22)	2.90	$50.00	$75.00	+((K10*M10)+(K10*O10))
11				@DATE(93,6,8)	4.00	$50.00	$75.00	+((K11*M11)+(K11*O11))
12				@DATE(93,7,1)	3.70	$50.00	$75.00	+((K12*M12)+(K12*O12))
13				@DATE(93,8,15)	3.50	$50.00	$75.00	+((K13*M13)+(K13*O13))
15				Mean:	@AVG(K8..K13)		Mean:	@AVG(Q8..Q13)
16				Std. Dev.:	@STD(K8..K13)			
17				Range:	+(@MAX(K8..K13)-@MIN(K8..K13))			

Figure 21-2. Formulas for MLT Data

of abstraction and may show only the relevant portions of the entire database. It follows that there may be many views of a database. An example of a database such as a small MRP system has views of BOM, inventory, general ledger, order entry, and purchasing.

The components of a database are attributes (also called fields or data items), entities (also called records), and entity sets. Attributes are specific individual entries (e.g., a company's name or a product name) and the entity (e.g., the entire name of the company, address, and phone number or product name and model numbers). A group of entities make up an entity set or file. A relationship links entities (e.g., a purchase order number links a customer's entity and the product names and model numbers for the parts that are ordered).

The three designs of databases are the object-based logical model, record-based logical model, and physical data model. There are a variety of object-based logical models, including the entity-relationship and object-oriented models. The entity-relationship model is similar to a person's perception of the real world. The attributes make up entities with relationships linking entities. The object-oriented model is based on object-oriented programming with the code and data combined into a unit called an object. Record-based logical models describe the data at the *view* and *conceptual levels*. The record-based model is designed on fixed format records with a fixed number of fixed length fields or attributes. Record-based logical models require separate languages to express queries and updates. Physical-model databases have their design transformed into data constructs defined by the database management system being used, and the design is matched into the constraints of the specific program. The physical design is database dependent. The common relational model database is a type of record-based logical model. Physical data models operate at the *physical level*.

Database schemes define the variables in a database. The physical—or lowest—level, the conceptual—or intermediate—level, and the subscheme—or highest—level are the three types.

Database models describe the structure and processing of a database. Database models are used for both logical schema and physical schema. The two models of databases are *data definition language* and the *data manipulation language*. The *data definition language* includes terms for the storage structure and access methods for the database. This is usually hidden from users. The *data manipulation language* describes the processing of the data by insertion, retrieval, deletion, and modification of data in the database. The *data manipulation language* enables a user to easily access the database. A query requests information from the database. A query is written in a *data manipulation language*.

Query languages, such as SQL, QBE, and Quel are high-level languages used to request information from databases. SQL is a user-friendly query language that includes both data definition and data manipulation features for queries of the relational model database (a type of record-based logical model). SQL is a transform-oriented relational language that transforms inputs into desired outputs via relations. SQL can be executed as a query and update language by itself, or it can be embedded in commands in application programs (See page 209).

The organization of a corporation's database(s) is a significant issue. One central database for a multiplant company means that everyone is working from the same database. This ensures compatibility of the database with corporate software. This also localizes the expertise required to manage the database, but it reduces flexibility and awareness of local issues. The centralized database does require consistent, reliable computer communications between the various corporate offices and plants. Distributed databases require more local computer and database management with opportunity for incompatibility between the various databases and application programs.

21.6 Machine Tool Programming

Computer numerical controls are outgrowths of the hard-wired numerical control systems developed in the 1950s. Today the terms NC and CNC are used interchangeably.

Numerical control is the operation of a machine through a tool path by a series of coded alphanumeric instructions.

Computer control of machines has changed manufacturing technology more than any other single development, introducing as it did the development of automatic machine control. In so doing, it broke the ground that enabled manufacturing cells, FMS, robots, coordinate measuring machines, and programmable logic controllers to be developed.

One might well add that the concept extends to very large scale integration (VLSI) which has helped provide more capable computer control units at a lower cost. This is parallel to the early machine tools that were fabricated to make leadscrews which helped build more accurate and capable machine tools.

21.7 CAD/CAM

Mechanical. Computer-aided design/computer-aided manufacturing (CAD/CAM) has become a significant factor in the design and manufacturing process. Once perceived only as a tool for Fortune 500 companies, CAD/CAM now has become affordable by all. Most CAD systems are designed and intended to automate manual functions, whether they be engineering analysis, conceptual design, drafting, or documentation.

Two Dimensions. While early CAD systems were 2-D, today most are 2½ or 3-D. Two-dimension replicates manual design, detailing, and drafting. Two-and-one-half-dimension CAD systems have knowledge of the third dimension and even the ability to display it without the computer overhead of storing the full geometric representation. Two-and-one-half-dimension CAD continues to play

SQL Keywords	SQL Functions	SQL Logical Operators
SELECT Retrieve existing data	COUNT SUM AVG	EXISTS NOT EXISTS
INSERT Inserting data	MAX MIN	
DELETE Deleting data		
UPDATE Modify SET existing data		
FROM The relation to be used.		
WHERE Can contain modifiers for delimiting the range of sortation		

```
    The syntax for a SQL command to change a part from one
machine (#1) to a different machine (#2) may be written as:

    UPDATE PART
    SET PART.MACHINE # = 2
      WHERE PART.MACHINE # = 1
```

a role in companies that deal largely in simple or purchased parts where interface, interaction, or interference between parts is specified in contrast to being a design function.

Good 2-D or 2½-D CAD software packages will incorporate the following attributes:

• Good line and text font capability,

• Full support of ANSI and ISO standard drawing sizes,

• High interactive response speeds,

• Tools for creating and storing libraries,

• Strong dimensioning package that conforms to the various standards,

• Tools supporting conventional and metric dimensioning units,

• Tools supporting geometric dimensioning and tolerancing units, and

• Entity drawing layering capability.

CAD software packages may have features that do not support the secondary or further use of the drawing data.

Questions to be answered should include:

- Is the system vector or raster based? Vector supports CAM and manufacturing applications best, and raster supports scanning, displaying, and editing paper drawings best.

- How are curves represented?

- Can the output be scaled?

- Is there an upgrade path to 3-D?

- Can the system be used to detail 3-D models?

Three Dimensions. Three-dimensional modeling gives a full and complete CAD/CAM representation. Three-dimensional CAD systems consist of wireframe, surface, and solid modeling systems. Their 3-D capability is necessary to both thoroughly describe all but the simplest of mechanical parts and to provide multi-axis computer numerical control of machine tools.

Wireframe consists of storing geometry of the 3-D model as edges and points. It is transparent in nature, requiring some skill and expertise in interpreting the model. The points on the surface are implied. Precise surface representations or information must be passed through line and offset-type information. Among the advantages of wireframe is the infinite number of views and drawings that can be generated from a single model. However, it is moderately difficult to clean up the drawing to make a finished engineering drawing.

Surface modeling adds varying degrees of accuracy in a CAD/CAM model when compared to wireframe models. Planar or ruled surfaces or surfaces of revolution have increased accuracy, while sculptured surfaces have lesser degrees of accuracy than wireframes. The additional surface information gives improved graphical imaging when it is linked to $2\frac{1}{2}$, 3, and 5-axis manufacturing applications and numerical control processes.

Solid modeling consists of constructive solid geometry (CSG) or boundary representation solids (BREP). Constructive solid geometry uses primitives (cubes, cylinders, cones, tori, etc.) to create solid images. In the CSG system, the solids are created by storing construction parameters and size to specified primitives. These primitives are combined to form a composite solid object. CSG is most appropriate for regular prismatic components, and, while limited, the number of primitives is increasing on many systems.

BREP solids can be stored as true surfaced or faceted. The true surface form is stored as true surface representation and topology. The faceted form stores the faceted surface representation and true surface data is generated when needed by other applications. Generally, the construction sequence of the BREP solid is stored with the associated BREP solid.

Solid modeling provides CAD/CAM systems with a wealth of knowledge. Topological models give CAD/CAM systems the capability of understanding the inside and the outside of a model, including maximum and minimum material conditions. Solid modeling requires more computer power than surfaces or wireframe, but drafting, engineering analysis, and CAM functions can be executed faster from solid models than the other systems.

Electrical. The capture and conversion of hardcopy electrical schematic and printed circuit board (PCB) drawings can be accomplished with a scanner and CAD software that supports raster and raster-to-vector transition. This permits existing electrical schematic and PCB drawings to be converted to CAD files and edited and replotted. This is fundamentally easier with electrical schematic and PCB drawings because they are strictly 2-dimensional drawings in contrast to 2-D object drawings.

Current technology enables the electrical/electronic circuit design and computer-aided engineering functions to include all tasks in electronics design (analysis, design, verification, and simulation). The design, analysis, and simulation capabilities of CAE systems is causing the electronics industry to move from CAD to CAE.

Printed Circuit Boards. CAD software designed for printed circuit boards (PCB) has features unique to that application. Current surface mount technology (SMT) and the continued miniaturizing of integrated circuit products makes the design of most PCBs a complex task. Contributing to the complexity of designing PCBs are:

- The number of layers in a final board assembly (single-sided, double-sided, or multilayered);

- The miniaturization of components and the effect on pin spacing and number of pins in a conductor;

- The need for filtering related to conductor placement (high-frequency switch causes inductance);

- Conductor routing and board layers;

- The frequency of the current in the different circuits and the resulting inductance;

- Heat dissipation;

- Capacitance of parallel conductors; and

- The placement of similar types of components.

PCB-oriented CAD software can assist in board layout and routing and in accommodating the above points versus attempting PCB design manually or with non-PCB CAD software. Manual layout gives greater flexibility in component placement and in determining if conductor placement is an inductive or capacitive issue, but it requires electrical expertise. For an automatic system to be useful, the rules constraining the layouts (e.g., board size, components, component placement, etc.) must be established. This may be time-consuming. The advantage is that as the rules are developed, they are done only once per type of board.

21.8 CAM/CIM

Manufacturing Networks. Manufacturing networks are medium- to high-speed networks operating within a limited geographical area called a local area network (LAN). Manufacturing enterprises have different requirements for computing, data, and information than do offices. Manufacturing LANs should, like any other automation, improve quality, improve production efficiency, improve quality of work life, and improve product development time. As with other types of automation, automation may be a corporation's requirement for their business plan or a customer's requirement for doing business with them. If a firm wishes to move toward computer-integrated manufacturing (CIM), networking is an absolute requirement. Many companies require their suppliers to be able to send and receive data via electronic data interchange (EDI) or modems. In plant, the network may be used to communicate CAD drawings, process engineering information, process routings, shop-floor production schedules, manufacturing programs (CNC, RLL, robotic, and others), shop-floor data or information, quality data, and inventory and material handling data.

The hardware and software of networks needs to be selected against a backdrop of distance spanned by the network, number of nodes or users on the network, data throughput rate, conductor characteristics, network protocol, installation costs, hardware and software compatibility for multi-vendor systems, interfacing versus integration, and the corporate goals set for the LAN. The issue of hardware compatibility and the interchange of information has been the driving force behind the development of open systems network architecture. Three common open systems found in manufacturing enterprises are the Technical Office Protocol (TOP) developed by Boeing for the technical office; Manufacturing Automation Protocol (MAP) built on the Open System Interconnection (OSI) reference model developed by the International Standards Organization (ISO); and Transport Control Protocol/Internet Protocol (TCP/IP) sponsored by the Department of Defense.

Smaller manufacturing companies often use proprietary network systems such as DECnet and Novell, owing largely to their not being able to afford MAP or open systems architecture. Many do not require it for in-house operations or they do not have the expertise to maintain it. Instead, they have installed networks proprietary to their operations. Yet, although these LANs have a large installed base, if used alone, they tend to result in interfacing and not integration.

Proprietary systems such as ethernet, DECnet, or Novell that have a large market share have features that give some of the capabilities of MAP/TOP at a fraction of the computing power, hardware and software, and technical expertise costs.

Enterprise-wide computing (EWC) and wide area network (WAN) computing may be similar in operation but different in organizational purpose and function. Both cover larger areas than LANs and typically include multiple LANs. WANs moreover are not necessarily limited to a single organization. The simplest type of WAN is one linked by modems. Faster and more reliable transmission of data becomes more expensive but may still use the telephone services.

Topologies. The three most common network topologies are star, token ring, and bus. A fourth can be created by way of hybrids of the these. In all but the simplest of installations, an actual network consists of combinations of bus or ring with stars clustered in different locations of a star, bus, or ring as needed by the physical and user parameters of the network. The selection of a topology is determined by the distance spanned by the network, the protocol used by the network operating system, the number of users, and the desired throughput on the network. Ethernet and ARCnet use bus, star, and hybrid topologies. IBM-token ring uses hybrid token ring with stars located on the ring. MAP uses a token passing bus.

Conductors. The four common conductors for networks are twisted pair (TWP), coaxial cable (*flexible and rigid*), and fiberoptic cable. Ethernet uses *thicknet* (rigid) or *thinnet* (flexible) coaxial cable, TWP wire (for 10Base-T), or fiberoptic cable. Ethernet systems using *thicknet* have a larger diameter coaxial cable or fiberoptic cable. This hardware can span a greater distance but is more costly and difficult to install or modify than *thinnet* and TWP. *Thinnet* uses bus, star, or hybrid bus topologies with coaxial cable or fiberoptic cable busses and star hubs with TWP (10Base-T) to nodes on the bus. Token ring uses flexible coaxial cable between nodes—multiple access units (MAU)—on the ring and between nodes on the ring and hubs. Between hubs and personal computers token ring may use flexible coaxial cable or TWP (10Base-T), depending on the network hardware.

Coaxial cable is more expensive but less susceptible to electrical noise than either shielded or unshielded twisted pair (STP, UTP). Unless a distance of a kilometer or more is covered or multiplexing takes place, fiberoptic cable may not be cost-effective. Fiberoptic cable is optimum for multiplexed signals because of the information capacity of the high-frequency signal. Signals on fiberoptic cables can be split, but the hardware adds to the cost. Twisted pair was most commonly used with ARCnet, but is now being used with ethernet and token ring with 10Base-T hardware. MAP uses rigid coaxial or fiberoptic cables. TCP/IP uses ethernet conductors.

Baseband. Baseband has a digital signal that (1) is not modulated by a carrier and (2) cannot be multiplexed. It typically uses TWP or flexible coaxial cable, although fiberoptic cable is technically possible. The entire capacity of the cable is used during each transmission. TCP\IP and ethernet, along with most other proprietary network operating systems for LANs, operate in baseband. Ethernet operates at 10 megabits per second, ARCnet is slower at 2.5 megabits per second, and IBM token ring transmits at either 4 or 20 megabits per second.

Broadband. Broadband has a high-frequency carrier-modulated signal that may be multiplexed. Because broadband systems have multiple channels, they require additional hardware to modulate and demodulate the signal and are typically more expensive than baseband systems, yet are generally more insulated from plant-floor electrical noise. Multiple channels enable broadband to carry computer-based data, video data as in closed-circuit television, and audio data for voice communications all on the same network. The broadband requires a transmitter/receiver to provide the carrier for the communications. MAP is the most notable user of broadband and transmits at 10 megabits per second.

Protocols. Protocol is the procedure by which a node on the network is allowed to access the network. The three most common protocols are carrier sense multiple access with collision detection (CSMA/CD), token ring, and token passing. Ethernet always uses CSMA/CD, IBM's token ring uses token ring, and ARCnet and MAP use token passing.

A node on the network wanting to communicate with the host or server by way of CSMA/CD (1) waits for a pause in the transmission of data on the network, (2) begins transmission, and (3) checks that no other node has tried to transmit at the same time. These three steps constitute "carrier sense multiple access." With steps (2) and (3), if another node tries to transmit on the network at the same time, a collision of data will occur, (4) the nodes will sense the collision and cease transmission, and (5) each node will wait a prescribed time based on a random number generated in the network interface card (NIC). The node then will check for a clear line (no transmissions) and begin to transmit its message again.

Token ring systems pass a signal to the next node on the ring. That node receives the token, address, and message, and if it is not for that node, it retransmits the message. This receiving/retransmitting continues until the message reaches the node or MAU of the appropriate address. If the node does not have a message to transmit, the token is released. Each MAU may be the hub for a star to individual PC nodes.

Open Systems Interconnection is the base for General Motors' Manufacturing Automation Protocol, or MAP. The MAP effort was an attempt to simplify the OSI standards with respect to the various options and provide for hardware and software compatibility within and outside of GM. MAP simplifies the options by describing all seven layers of the protocol stack with greater definition than did the OSI model. Map 3.0 is now more than five years old, but has yet to fulfill the goal of being a simple-to-integrate system. It commands a huge amount of computing power and is expensive, but it still offers the most integrated system for large manufacturing operations. It holds great potential for companies spread over a large geographical area, with large amounts of data to be transferred, and with a large number of users and diverse hardware. While MAP is certainly alive, the other technologies of TCP/IP, ethernet, and IBM token ring have their user base, with ethernet having the largest factory-floor user base. In many of these installations, nodes are networked or interfaced but not integrated.

Fundamental Concepts. Zero-LAN, peer-to-peer, client-server, and host-terminal are common LAN configurations. Zero-slot LANs and peer-to-peer LANs lack security and are slow, cumbersome, and difficult to manage. Client-server LANs are microcomputer- or minicomputer-based file servers with microcomputers or engineering workstations on the network. The microcomputers are capable of executing the network application programs locally and storing the data either locally or on the file server.

The host-terminal configuration is a minicomputer or mainframe computer host with dumb, smart, or intelligent terminals. All of the processing is done on the mini- or mainframe computer. When the host has multiple file servers, the hosts are called clients and a client-server relationship exists with the hosts as clients accessing a set of servers. It is possible to use a microcomputer in an emulation mode in which the microcomputer emulates a dumb, smart, or intelligent terminal and communicates with the host in that manner. In this mode, communications software may be used to transfer files from the mainframe or minicomputer to a microcomputer file. If the nodes are diskless, the system has better file security.

Network Operating Systems. The management of the file server, files, system faults, security, client or terminal access to the system or files, protocol, printing, and communicating with other server or host systems is the role of the network operating system (NOS).

Bridges, Routers, and Gateways. Repeaters and bridges are similar in application but different in capability. Repeaters operate at the ISO model physical level (Layer 1) and communicate in bits. They typically do not do error control, flow control, or address correction. Repeaters are reasonably fast because they do not process the signal.

Bridges are used to link identical LANs to increase user access to greater range. This linkage generally increases user access to file servers and application software, number of users and e-mail communications, and printing resources. Bridges store and forward frames at the data link level (layer 2) of the ISO model.

Routers are protocol-sensitive units that support communication between dissimilar LANs (architectures) using the same protocol. Novell's SPX/IPX or DEC's DECnet are examples of common protocols for ethernet and token ring architectures. Routers operate at the third level of the ISO model (network layer) and communicate in packets.

Gateways connect networks of different architectures and protocols by translating the protocol from one to the other. They are used between minis and mainframes, LANs and WANs, and LANs of different architectures and protocols. They process bits at the physical layer all through error detection, framing, routing, flow control, message byte field interpretation, etc. at the appropriate level from the physical

through the application layers. The advantage of being able to translate and connect a network to any other network has the disadvantage of a time delay in the propagation of the message. Repeaters and bridges are much faster, but they are less capable and have less data to process.

Programmable Logic Controllers. Programmable logic controllers (PLCs) are one of the primary forms of manufacturing automation. They are used as relay replacement, control of analog and digital closed-loop systems, and manufacturing cell control in three levels of manufacturing automation.

The use of PLCs as relay replacement is one of their major functions. This application results in PLCs of all sizes being used to replace existing relay control panels or to control new equipment. Several advantages arise from such replacement: ease of program modification at later dates, capability of collecting data relevant to the process under control, communication with other equipment at the bit input/output (I/O) level, communication with other computer equipment at either serial or network levels, and troubleshooting the control system. However, if PLCs without data communication capability are selected, this application results in an "island of automation"; that is, an automated system for that process with minimal capability (I/O bits) of communicating with the remainder of the manufacturing operations. PLCs have a maximum current-carrying capability of two amperes. It is common to find one or even two relays between a PLC and a large motor starter because of current incompatibilities.

PLCs used in analog and digital closed-loop control systems are quite common with the migration of PLCs to more powerful central processing units. Historically, analog inputs and outputs were cumbersome to program. Because of the increase in variety and sophistication of programming languages and embedded commands, the difficulty in programming analog functions has been reduced.

Cell control can be accomplished by a computer-based cell controller, but PLCs are more common for this function. Even when a piece of automated equipment (e.g., a robot controller) has the capability of performing cell control, manufacturing and process engineers should consider using PLCs to interface with the robot to control the manufacturing cell because several advantages accrue:

- Increased control capability (approximating co-processing);

- Faster processing of the robotic program without the additional cell control statements;

Simpler robotic programs; and

In the event the robot controller is not functioning, the material handling or process functions of the robot may be performed in some other manner and the cell will still be controlled by the PLC.

Likewise, if the PLC has a failure, the diagnosis is easier and the time-to-repair is reduced.

Programming. PLCs are programmed primarily in relay ladder logic. This language is optimum for factory electricians and relay replacement but is weak in higher level mathematical, analog, and servo functions. Many PLCs can now be programmed in Boolean algebra, BASIC, C, or a sequential function language which permits the higher level functions to be programmed more easily.

The basic elements in a relay ladder logic (RLL) program are:

- RAILS,
- RUNGS,
- BRANCHES,
- INPUTS [EXAMINE ON input ($-| \;\; |-$) and EXAMINE OFF ($-| / |-$)], and
- OUTPUTS [OUTPUT ($-(\;\;)-$), TIMER ($-[T]-$), COUNTER ($-[CNT]-$), sometimes ($-[CTR]-$)

RAILS are vertical lines serving as the source of EMF for relay circuits and logic for the PLC. RUNGS are horizontal and contain the BRANCHES, INPUTS, and OUTPUTS. A BRANCH starts and ends an OR function. The INPUT, EXAMINE ON is true only when the input is HIGH or ON. The INPUT, EXAMINE OFF is true only when the input is LOW or OFF. EXAMINE OFF performs a NOT logic function. The OUTPUT, a logic bit in memory as well as on the OUTPUT card, is sometimes called a coil in reference to its predecessor relay coil. It is generally on the right-hand side of the rung.

Most PLC manufacturers use microcomputers for programming the PLC, in place of the traditional dedicated programming terminal. Programs can be stored on disk instead of the older, slower cassette tape storage media.

Logic. The logic in a PLC consists of OR, AND, and NOT Boolean functions. Other standard features are $=$, $>$, $<$, timing, counting, sequential, arithmetic functions, matrix, and word and block shifts or transfers. Higher level optional functions are servo, analog input, analog output, and PID control.

Fundamental Concepts. The basic blocks of a PLC are similar to those of a microcomputer. They have a power supply, central processing unit (CPU), memory (RAM), operating system, and input/output modules. The operating system for the PLC is designed to permit the programmer to easily access the input/output bits in a timely manner, a task that is not so simple with a PC. The input and output cards identify terminals so they have the same state as the binary input or output memory table (ON, 1, TRUE, or OFF, 0, or FALSE).

Owing to the nature of the hardware a PLC controls, safety considerations mandate local control of a system for start-up, emergency shutdown, and lockout. Further safety considerations require that all EMERGENCY STOPS be hardwired and not wired as inputs into the PLC control system. When a PC fails, data is lost. When a PLC fails, dies, products, and even the operator can be lost. Despite the efforts of PLC manufacturers, it is impossible to accurately predict how a PLC will behave if the CPU, memory, or I/O cards fail. It is recommended that hardwired closed-loop control circuitry be incorporated to stop a machine if the PLC fails.

Computer Numerical Control. CNC, as a component of CIM, is often used in a standalone mode, thereby becoming an island of automation. To gain the maximum use of the flexibility in scheduling these machines requires that, at a minimum, they be interfaced with a computer network to permit the transfer of production scheduling information, programs, and operator and job information. This interfacing is a step up from an isolated island and provides the integration necessary for CIM or FMS provided with distributed numerical control.

Direct Numerical Control. In direct numerical control systems, a computer with part programs is linked to NC machines. The computer downloads the part programs to NC machines a block at a time (just as the NC machine runs paper tape a block at a time for each part) in a time-share mode with the other NC machines linked to the computer. Upgrades to numerical control systems were to have the advantage of not having to manage all of the rolls of paper tape that had to be executed a block at a time for every part. However, these upgrades and improvements were even harder to manage. Direct numerical control has evolved into distributed numerical control with the advent of CNC and the expanded capability of the local CNC machine to hold and execute entire part programs.

Distributed numerical control holds part programs in a central location and allows a CNC machine controller to request (1) a part program and (2) that the program be downloaded. Typically, a distributed numerical control system will ensure accurate transmission of data in the part program. It will also pass data from the CNC to the host computer regarding program enhancements and production piece counts, quality control information, machine downtime, and tooling information. Distributed numerical control has emerged as a major building block of a CIM system.

Control Systems Theory. The primary elements of a CNC machine control unit are an operator interface, machine control unit, and machine interface. Common operator interfaces are keyboards, papertape punch/readers, magnetic tape readers, host computers, and network interfaces (e.g., RS-232C, EIA-232D, RS-422, ethernet, etc.). The machine control unit, or CPU, has all the characteristics of a microcomputer: memory, both read only (ROM) and random access (RAM); an arithmetic unit; and a control unit.

The machine interface passes outputs and inputs between the control unit and the processing machine. The machine interface transforms the digital output signal ($\pm 5v$) of the control into an amplified analog AC or DC signal or an amplified pulse-width modulated DC signal to drive the motors on the processing machine. It also transforms closed-loop feedback generated by resolvers or encoders from the motors or ball leadscrews. Other outputs include discrete signals for coolant pump operation; powered tools mounted in turrets on turning centers; and hydraulic valve operation for power clamping, indexing tables, power chucks, and power tailstocks. Input from the machine to the control unit to give closed-loop feedback is generated by resolvers or encoders mounted on the servo motors or ball leadscrews that indicate the commanded position has been achieved. Additional discrete inputs include:

- Proximity switches indicating extreme limit and HOME locations for each axis;

- Safety mechanical switches for cabinet and enclosure doors;

- A pressure switch or oil level switch for lubrication of the spindle and ways;

- An air-pressure switch for machine tool change operation; and,

- Heat-sensitive switches for cabinet, oil, and motor temperatures.

Most control interfaces have additional discrete I/O for linking the CNC machine to pallet changers, programmable logic controllers, or cell controllers.

The motion control for CNC is open-loop stepping-motor or closed-loop servo-control. In general, open-loop control systems generate commands based on time, discrete on/off control, or steps for stepping motors. Motion control systems receive no feedback information or signals to indicate if the motion command was completed. Open-loop control is akin to that of older garage door openers, plotters, most printers, and most stepper motor drives. Its obvious weakness is its inability to detect if the motion control caused the mechanism to reach the desired objective. Open-loop control is typically found on the lower end of the equipment cost scale.

Closed-loop control is used with discrete systems as well as servo systems. It generates information called feedback to the control unit that the controller compares with the motion command to determine if the motion command was achieved. CNC equipment, robots, and cruise control on automobiles are common closed-loop systems. PLC-controlled fluid power valves that control cylinders with limit switches are a discrete closed-loop system. The limit switches provide inputs indicating that the motion of the cylinder was completed. Cabinet door or guard limit switches that prevent machines from operating unless the door or guard is secured are closed-loop.

CNC machine control units may have fixed or floating zero points. A CNC machine with a fixed zero point has the zero-zero location (origin) of the table at a specific location, typically the left front edge. The programmer must locate all coordinates in a positive x-y from that location. The more common floating zero system allows the operator to determine where the zero (origin) point should be set. The floating point is a much more convenient system for the CNC user.

Programming Formats. The standard programming instructions include the tool path turning the spindle on/off, coolant on/off, tool changes, and machine feeds and speeds. Additional instructions for specific CNC tools (e.g., turning centers and grinding operations) are common. A CNC machine consists of a program of instructions, the machine control unit, and the processing hardware. The common format for program entry in early NC machines was paper tape; today, however, paper tape provides hardcopy backup for most CNC users.

The standard G & M codes for CNC are listed in Appendix C of this book. The syntax for the code is block number; G instruction(s), and the X, Y, Z, a, b, c, j, k, l parameters with numerical values; and decimal points for the instruction. These are followed by the spindle speed and feed rate if the speed or feed is changed from the value in the previous block. Spindle operation, coolant operation, tool changes for machining centers or chuck operation, and tailstock operations for lathes or turning center lathes are done in M codes. Controllers of different manufacturers may have some differences in syntax for executing the program. These programs may be run in single-step or automatic modes. While automatic is used for production, one generally uses single-block for checking or editing a program. Sample turning and milling programs with explanations are shown in Figures 21-3 and 21-4.

linear and circular interpolations in the CNC operating systems and the software may incorporate helical, parabolic, and cubic interpolations. These interpolations eliminate a large number of data points between the starting and specified points and thus achieve a level of accuracy of machine movement that is not practical without interpolation.

The earliest NC language (computer-assisted programming) was Automatic Programming of Tools (APT). This language is still popular with companies doing 4- or 5-axis

Program	Explanation
N0740 G50 S1000	Button tool
N0750 G20 G96 S1000 M4	Limit spindle speed
N0760 G00 Z3.487 T0101	Rapid to Z-position
N0765 G00 X.6	Rapid to X-position
N0770 G42 G00 W-.746	Tool compensation, Rapid to Z-axis position
N0780 G20 G98 F3.0	Feed definition
N0790 G01 X.370	Feed X-axis
N0800 G01 W-.240	Feed Z-axis
N0810 G01 X.570 W-.0300	Feed
N0820 G00 X.7	Rapid out
N0880 G28 U0.0 W0.0 T0100	
N0890 T0800	

Figure 21-3. Sample Turning Program

Program	Explanation
N160G00X-3.Y-3.T4M6	Tool change, Tool #6
N170G00G90X-1.755Y-1.062Z-.075	Rapid traverse
N180G01Z-.188F160S4000	Feed Z-axis
N190G01X1.755	Feed X-axis
N210G03X1.937Y-.88I1.755J-.88F160S4000	Counterclockwise circular interpolation
N220G01Y.88	Feed Y-axis
N230G03X1.755Y1.062I1.755J.88F160S4000	Counterclockwise circular interpolation
N240G01X-1.755	Feed X-axis
N250G03X-1.937Y.88I-1.755J.88F160S4000	Counterclockwise circular interpolation
N260G01Y-.88F160S4000	Feed Y-axis
N270G03X-1.755Y-1.062I-1.755J-.88F160S4000	Counterclockwise circular interpolation
N280G00Z-.07	Rapid Z, up
N430G00G90X-3.Y-3.T1M6	Tool change, Tool #1

Figure 21-4. Sample Milling Program

In most companies, the preparation of CNC programs for all but the simplest of geometric parts is done with computer-aided manufacturing (CAM) software. CAM systems have

CNC operations, especially airframe and machine tool builders. Other languages that have seen popularity in the last 30 years are Compact, Split, and Autospot. With the exception of APT, these languages, although still used, are not nearly as popular in light of the advent of 3-D CAD/CAM packages.

Computer-assisted part programming included the operating system specific to each proprietary language, the compiler, the NC processor, and the post-processor. The NC processor generates an intermediate file called a *cutter location file* that contains cutter location data. The post-processor is specific for each brand of machine control unit and converts the cutter location data into a tape file for that specific machine tool.

Standard Codes. The evolution of standards for paper tape has resulted in two common forms: EIA-244B and EIA-358B. Both are eight-channel one-inch tape. Tape form EIA-244B uses channel five for parity (odd) and channel eight for end-of-block, while EIA-358B uses channel eight for parity (even). CNC specialists need to be familiar with these two standards for two reasons: first, when uploading or downloading programs electronically, they need to know the parity along with other communication parameters; second, if they are to use paper tape, the formats must be compatible. While current controllers may be set to accept both formats, compatible formats are necessary for program exchange or DNC configuration.

Machine Coordinate Axes. A CNC machine tool's linear axes start with Z (upper case) and move towards A in the alphabet with one letter per axis. The Z axis is always parallel to the spindle, X is always perpendicular to Z and is the major axis on the machine, and Y is perpendicular to both X and Z and has a smaller movement than the X axis. If a CNC machine has multiple spindles, U, V, and W—parallel to X, Y, and Z, respectively—are used for the second spindle/table/turret configuration. When a machine has rotation about X, Y, or Z, the rotation is designated by lower case a, b, and c, respectively. Two-axis systems (lathes and turning centers) are X and Z. Three-axis systems (milling machines and vertical and horizontal machining centers) are X, Y, and Z. The fourth and fifth axes (rotations about X and Y) are a or b, respectively.

The right-hand rule (Figure 21-5) is helpful for keeping axes in the correct orientation. The thumb points toward +X, the index (first) finger is perpendicular to the thumb and points toward +Y and the second finger is perpendicular to both the thumb and the index finger and points toward the +Z axis.

Transfer Media. The traditional transfer media for part programs were paper tape; magnetic tape, disk, or drum; operator station; punched cards; host computer; and

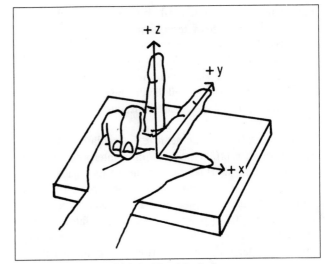

Figure 21-5. The "Right-Hand Rule."

modems. Current common technology uses operator station, disk, host computer, or network (ethernet) interface. The operator station and manual data input is used because current controllers enable the operator to enter a program while the controller is executing an active program. Floppy disks used in environmentally hardened disk drives are available on CNC control units. Host computers typically are connected to the control unit via RS-232C, EIA-232D, or RS-422. Ethernet connections have been requested by users because of the large use of VAX systems (DECNet) on the factory floor.

Data Representation. All data in CNC is represented in binary. Each row of a paper tape is a single alphanumeric character. The code transferred over the serial communications line with a host computer is in ASCII conforming to RS-232C (EIA-232D). Since NC code does not recognize all alphanumeric characters, six bits plus parity and end-of-block bits are adequate to describe the standard 63 (2^6) characters in CNC.

Point-to-point Operations. All early NC machines were point-to-point units which moved either the tool (lathe/turning center) or table/workpiece (drilling) to a given location. The control unit in these machines attempts to move each axis independently of each other and the tool path can only be approximated. Current control system technology for turning and machining centers have both linear and circular interpolation. In a literal sense, CNC drilling and tapping machines do not require more than point-to-point operation because that is the function of the tool.

Interpolation. Linear and circular interpolations are standard features on many CNC machines. These features provide the capability of moving multiple axes in simultaneous control to generate a 3-D line or a 2-D circle (in the principal plane). The interpolator may be considered a small, fixed program computer. It receives slide direction and measurement calculations from active storage and directions as to how fast the cut of the path is to be made. It then calculates the data and directs the movement of each slide at the correct time/distant constants.

Linear interpolations can achieve interpolations to increments as small as 0.0001 inch. A linear interpolation can be programmed as short as the tolerances the machine will allow. With CAM software, complex and free-form curves can be accurately estimated by the computer, generating a very large number of points of linear interpolated points to estimate the curve. Circular interpolation is the most common higher order interpolation (parabolic and helical being others). It is used to approximate circles in the principal plane (2-D) of the machine. Parabolic interpolation is a second higher order interpolation. Though it is not as effective with circles as circular interpolation, it is generally more efficient with other curves.

Contouring Operations. CNC mills, lathes, machining centers, turning centers, and grinding machines can generate complex shapes. Limited-contouring machine control systems have been built with circular interpolation capability but without buffer storage. These machines were built to provide lower-cost CNC machine tools. Without the buffer storage, the control unit causes the machine tool to accelerate, machine, decelerate, and read in a new block. During the read time, the feed rate is greatly reduced, resulting in an average feed rate that is less than programmed. The reduced feed rate may also result in pulsating and undercuts. Full contouring machines have both interpolation and buffer storage. The buffer storage permits the tape reader to read the next block(s) and store the new data in the buffer. When the machine has reached the coordinates specified in the current block, the active storage data is dumped, the buffer downloads its data to the active storage, and the machine continues its interpolated motion at the specified feed. The tape reader reads in new block(s) and stores data for the next block in the buffer, while the active storage in the control unit executes the current interpolations. These data transfers require only microseconds, and the inertia of the mechanical system is sufficient to continue at the same velocity for the transfer time.

21.9 Material Handling/Automatic Identification

The basic elements of material handling include the motion of materials, the time required for moving the materials, the quantity of materials, and the space in which the materials is moved. Material handling needs to be considered as an integral component in the design of a manufacturing system, since it has an impact on the quality of the product, facility requirements, equipment layout, safety, productivity, and the physical integration of the manufacturing operation.

The two types of loads in material handling are unit and bulk. Unit loads are either single large pieces; parts or products that may or may not be placed on a pallet; or a quantity of smaller pieces, parts, or products that are combined onto a pallet or into a tote. Bulk materials are handled by conveying systems, bins, hoppers, and silos. The focus of this discussion is on unit load material handling systems.

Symptoms of inefficient material handling systems include disorganized storage (a lack of cube storage, material stored in aisles, poor housekeeping, overcrowding, and excessive handling); fragmented operations (built-in hinderance to flow, long hauls, excessive walking, and repetitive handling); and inefficient operations (lack of parts and supplies, inefficient use of labor, two-man lifting jobs, and trucks delayed or tied up).

Figure 21-6 details a structured seven-step process for planning material handling systems so they are an integrated part of the manufacturing operation. The seven steps are segmentation, flow development, alternative technologies, evaluation of performance of technologies, integration of subsystems, evaluation of costs, and plan of implementation for material handling.

Segmentation is dividing the facility into operations that require or interface with material handling. Receiving, storage, process workstations, packaging, and kitting are the fundamental elements of material handling. The information flow related to material handling consists of data entry and reporting for each of the segments.

Flow development, for both material and data, can be accomplished only after a clear and complete understanding of the functional operation and requirements of the subsystem. Operation process and flow process charts are useful in understanding the operations of the segment. Load movements are planned with *from-to* analysis. Considerations of volumes transported, frequencies of transport, the unit envelope, and the characteristics of the part/product being moved are additional considerations that need to be analyzed along with the *from-to* analysis. *Load-distance* and *activity-relationship* charts are additional analysis techniques. The data flow system provides information regard-

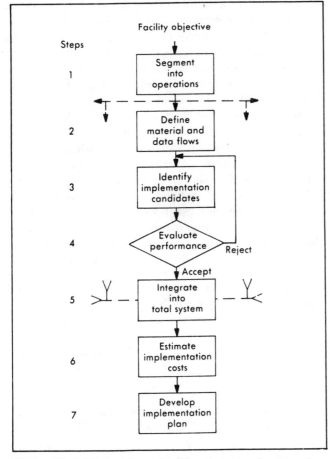

Figure 21-6. Methodology for Developing an Integrated Material handling System.

ing inventory, and material dispatch and tracking. This system may draw from the automatic identification, networking, and systems integration sections of the study guide.

Alternate technologies include potential candidate technologies that are considered for evaluation. Multiple factors to be considered when selecting technologies for evaluation include local configuration, orientation, and delivery frequencies. All of these can restrict traffic because of throughput restrictions at the load/unload stations. Process requirements for precise locations of parts can also influence the choice of technology. Facility constraints can also affect the technologies chosen for evaluation. For example, low ceilings may not permit overhead power-and-free conveyors or automated monorail equipment or concrete floors may not allow for in-floor conveyors.

Evaluation of performance of the alternate technologies may be performed with simulation software to verify the performance of a specific technology for a given subsystem (segment) for the technology under consideration. Once the

performance of a technology is verified, cost effectiveness assessments can be completed. This process should be completed for each of the appropriate technologies leading to the selection of the technology system meeting the performance and cost criteria.

Integration of subsystems occurs following evaluation of the interface of the subsystem technologies. The physical orientation, positioning, transfer times, and load transfer mechanism must all be considered. The information or data interfaces must also be taken into account at the subsystem interface. Appropriate and timely information must be provided. Once the physical and information systems have been integrated, the overall performance of the integrated system should be verified through simulation. This should be done for each additional step of integrating an additional subsystem. When all the subsystems have been integrated into physical and informational systems, the overall estimate of costs should be evaluated.

Costs of the material handling system include both the physical and information components. These consist of capital equipment costs for hardware and software, installation costs for hardware and software, training of personnel on both hardware and software, and supervision of the installation of the hardware and software. When the cost effectiveness of the system has been verified, implementation may proceed.

The implementation plan should be developed to include an installation and testing phased approach. The plan should enable a smooth transition into full system operation, reduce risks, and provide early phased payback from the portions of the system installed early in the project. Implementation should be supervised by experienced professionals.

Most material handling equipment is capable of moving unit loads. If the individual parts are small, they are combined on pallets or in a different container for a unit load. Pallet jacks, fork trucks, belt or roller conveyors, synchronized transfer units, and cranes are the typical and most common manual material handling equipment. Power and free conveyors, programmable monorails, automated storage and retrieval systems, robots, and automatic guided vehicles make up the common automated material handling systems.

Automatic Identification. Automatic identification improves both the accuracy (quality) and speed (production efficiency) of data entry for inventory, work in process, and final products. Bar code, radio frequency data transmission, radio frequency identification, magnetic stripe, voice recognition, and machine vision are the common forms of automatic identification used on the manufacturing floor. All of these systems are based on computer technology. Voice recognition and machine vision are both computer-intensive technologies that are being pushed to operate in faster and faster real-time applications. As computing power has increased so have the applications of the technologies. Various other forms of automatic identification such as optical character recognition are not common in manufacturing.

Bar codes. Bar coding is one of the most robust and reliable forms of automatic identification. Its error rate is approximately 1/10,000 of keyboard strokes. Bar codes come in multiple symbols and densities. Most common symbologies are Universal Product Code (UPC), Code 39, Interleaved 3 of 5, Code 128, etc. UPC is strictly numeric, while Code 39 is alphanumeric. The density of the bar code is high, medium, and low. High density, defined by the dimension of the narrowest bar (X dimension), is < than 0.010 inch, medium density occurs when 0.01 inch < X < 0.030 inch, and low density has an X value < 0.030 inch read from either side. Bar codes with $X \leq 0.020$ inch the wide dimension is specified to be 2.2 to 3 times the X dimension. When $X \leq 0.020$ inch, the wide dimension must be between two to three times the X dimension. Regardless of the density and the wide/narrow ratio, widths must be uniform throughout the code. In addition there is a nine-segment START and a nine-segment STOP bit on each end of the code. Each bar code must have a plain (quiet) zone on each end of the code.

Bar code is read from either end and the software adjusts for the direction of reading. The code is read by contact or non-contact fixed or non-contact rotating scanner. Rotating scanners are capable of making many scans per second.

Radio Frequency Data Transmission. Bar code data can be transmitted by radio frequency terminals to a host computer system. The radio frequency transmitter provides mobile real-time input of bar code data to the host computer and is under the control of the computer. The mobility is valuable around shipping/receiving docks, warehouses, and work-in-process storage. Advantages of this type of system include the direct, real-time transmission of data; broad access to information on any products in the host computer; and savings in labor stemming from the real-time access to data feature. The limited range of about a mile is the major limitation of radio frequency transmission.

Radio Frequency Identification. Radio frequency identification (RFI) utilizes an electronic, battery-powered unit called a transponder (tag) attached to a pallet or part to transmit information about the pallet number, parts, and manufacturing operations to be performed on the parts on the pallet. The responder (antenna) receives the data and communicates it to the computer system and the software. The data contained on the tag has 20 or more characters. These electronic tags are hardened to withstand vibration, liquids common in manufacturing, and temperature extremes of -40°F to 400°F (-4°C to 205°C). The electronic power of a tag ranges from milliwatts to 10 watts. As with all electronics, the tags are becoming smaller, more economical, and capable of transmitting and receiving more data. Since RFI is more costly than bar codes, its application must justify the additional cost. RFI is typically used in applications such as those in which a variety of parts with different manufacturing routings and processes may cover or destroy a bar code (e.g., painting). Current tags can receive and store as well as transmit (read/write) data.

Magnetic Stripe. Magnetic stripes, similar to those on bank and credit cards, are attached to pallets or parts. They contain more information and the information can be altered. Because they are more expensive, cannot be read remotely, and are not as resistant to mechanical damage, magnetic stripes are not as popular on the factory floor as bar code and are decreasing in use.

Voice Recognition. Voice recognition continues to evolve as a technology commanding attention, as the increased capability of digital computer systems has expanded the attributes of voice recognition technology. Application of the technology typically separates into two distinct groups: those with large vocabularies who access the systems in small numbers, and those with small vocabularies who access the system in large numbers.

In the group with larger vocabulary systems and limited user access, users typically record their pronunciation to establish a master pattern for their voice. When each individual logs onto the computer system at the beginning of a work shift, their voice patterns are verified by way of a few key words before they go on line. Typical applications are in situations where workers needed to have free hands, other auto-ID technologies were not applicable, and performance could justify the cost. The primary applications for voice recognition in manufacturing have been in material handling and inspection. Both may require the operator's hands to be free to hold or manipulate the part or inspection tools and the parts do not have bar codes.

Machine Vision. Machine vision is defined by the stages of (1) image formation, (2) image preprocessing, (3) image analysis, and (4) image interpretation. Systems that do not do all four of these, such as photosensitive switches or measuring systems, are not part of machine vision because they do not perform image analysis and interpretation. Image formation includes the optics and lighting to detect the image. Lens focal length, depth of field, aperture, and magnification are important optical parameters. Lighting is significant to the image formation because it helps develop contrast and a quality image. Image preprocessing receives the formed images every 17 to 33 milliseconds (60 to 30 times per second, respectively) the image is captured (frame and hold), converted from analog to digital values (analog to digital conversion (ADC)), and the digital image signal is stored in a frame buffer until it is processed. The analog values have a voltage value per picture element (pixel). The digital values are analyzed in terms of binary (black or white, digital zero, or one), gray scale (shades of black and white similar to black and white pictures using four, six, or eight bits of memory), or color. The image analysis is performed in a computer central processing unit (CPU). Such analysis is completed by describing and measuring the properties of several image features that consist of a region or a part of a region. Software techniques are available to analyze image features for position, geometric features (e.g., perimeter, diameter, centroid, area, or curvatures), or light intensity over the surface.

The comparison of features may be adequate to discriminate between parts, inspect parts for critical dimensions, or modify robot arm movements. Feature weighing is an additional important technique used in machine vision for measuring a feature and increasing the level of analysis. These frames may be transferred to robots for shifting a robot coordinate system according to a new part location. The frames may also be used in robot guidance.

Template matching has a "go/no-go" inspection function. The software develops a mask that matches a standard image. The "goodness of fit" between a frame and a template of the master image is the percentage of the number of pixels of the frame that match the number of pixels of the master template. A threshold "pass-fail" value can be assigned for accept-reject of a part. A probability of an incorrect go/no-go decision can be assigned that threshold.

Robotics. Robots, as defined by the Robotics International Association (RIA), are "reprogrammable, multifunctional manipulators designed to move material, parts, tools, or specialized devices through variable programmed motions...." Robots are composed of the four basic subsystems of all automation: power system, control system, robot arm (mechanical manipulator), and the robot arm/world interface.

Electricity is the power system of choice for robot arms. It is clean, quiet, capable of fast moves, well understood, and requires little maintenance. Hydraulic systems are used with larger payloads and slower moving robots. Both electric and hydraulic robots have servo control. Pneumatic power systems are restricted to non-servo control units and make up only a small percentage of the robot market.

The robot control system stores and executes programs, collects and records data, and communicates with other hardware or systems through discrete inputs/outputs and serial or parallel computer communications protocols. Most robots are still programmed with teach pendants. Where program changes are common and production should not be halted for programming, off-line programming can be used. Current control systems can be programmed in world coordinate systems, part coordinate systems, or tool coordinate systems. The world coordinate system is common for off-line programming. The tool coordinate system is used with arc welding where the angle of the tool (welding gun) is critical to the process.

Controllers are classified by their type of control, non-servo or servo; the resolution and accuracy of their control; the types of movement; and communications capability. Non-servo controllers may be a programmable logic controller and control robot arms that have limited positioning capability (low resolution) powered by pneumatic, hydraulic, or electric systems providing two to three positions per degree of freedom (DOF). Servo controllers are capable of high resolution and accuracy in position and velocity. Robot-controlled movement criteria are the number of axes they can control simultaneously, linear and/or circular interpolation, and continuous-path operation versus point-to-point. Communications capabilities include discrete input/output at the lowest type, ability to transfer programs over serial communications lines (RS-232C or RS-422), and the ability of programs to input (receive or transmit) data or frame shifts from machine vision systems to operating programs.

The robot arm (mechanical manipulator) is composed of links (solid structures) and joints to provide movement between links. Four types of joints are used: linear transverse, linear telescoping, rotary hinge, and rotary pivot or twist, and each joint is called a degree of freedom. Robot arms are classified by their work envelopes: cartesian and gantry robots have a cartesian work envelope, cylindrical and SCARA (selective compliance assembly robotic arm) have cylindrical work envelopes, and polar and jointed-arm (articulated) robot arms describe a spherical work envelope. The volume of the jointed arm spherical work envelope is significantly larger for the size of the robot than any of the other configurations.

The robot arm/world interface is custom tooled to interface the generic robot arm with a specific payload or process. If the process is resistance welding, spray painting, grinding/deburring, sealant/adhesive dispensing, assembly of mechanical or surface mount technology electrical component, or some other process, the arm/world interface is specific to the process. If the robot is grasping a part, the end-of-arm tooling must be designed to securely grasp the part, even with acceleration of multiple g's. The grasping is done with two- or three-fingered rigid grippers or three-fingered articulated grippers. If an application requires only grasping or hooking, a two-fingered rigid gripper is adequate. If other types of grasping are necessary, spherical, for example, articulated grippers are necessary. Sheet stock or parts with appropriate surface/volume ratios may be moved with either vacuum and suckers or electromagnets.

Robots are used to complete processes that add value to the product, will generally improve quality, production efficiency, and in some cases (e.g., painting, arc welding, or grinding/deburring) the quality of work life. These robots will make the greatest impact on manufacturing and will be the easiest to cost-justify.

In the first two decades of robotics, developers advocated the general purpose robot for all tasks. Today, the robots that have the best process performance are designed specifically for the process. Several types of equipment have robotic features but have become so specialized they are not technically robotic. Certain electronic surface-mount technology insertion machines and robot-controlled press brakes fall into this category. These specialized machines may compete with robot-based work cells.

FMS and FMC. In the 1980s flexible manufacturing systems (FMS) were projected as the solution to the nation's manufacturing problems. These highly sophisticated systems lost some of their attractiveness in the 1990s, but have now regained some of their stature. FMSs are a system of computer-controlled machines, usually CNC, with an integrated material handling system to move parts from machine to machine. The purpose of the FMS is to produce a family of parts or multiple part families in midvolume quantities. Flexible manufacturing cells (FMC), in contrast, have fewer machines and may not include the material handling function.

When FMCs are applied appropriately, they reduce the manufacturing lead time and the work in process. With a number of machines in the cell, production control is simplified and material movement is simplified and reduced. FMCs and statistical process control (SPC) complement each other. With the parts moving from machine to machine in a cell, a part that does not meet specification should be identified immediately. The FMC can accom-

plish its flexibility with manual intervention, in contrast to the complex and sophisticated software of an FMS. By definition and in practice, FMCs are less costly and simpler than FMS. This simplicity permits incremental implementation, with the option of integrating FMCs into an FMS.

Traditionally, midvolume, midvariety manufacturing processes were too small for continuous production and too large for batch production. Production efficiency, manufacturing lead time, and costs were problems. FMS with minimal setup and material handling time are theoretically able to economically produce a batch size of one. FMSs are generally recommended for family-of-parts processing in the midvolume range (1000-100,000), with a variety of processing operations. Direct labor is reduced, but indirect labor to set up, maintain, and program is increased from conventional batch manufacturing.

FMSs were viewed as the ultimate solution to problems confronting manufacturing in the 1980s, but many manufacturers using FMSs in the U.S. were never able to achieve the desired results for the investment. Owing to the complexity of hardware, software, and automated material handling of families of parts, successful implementation of FMSs has been an elusive target. Many FMSs are difficult to manage and do not produce quality parts at the desired rate with planned indirect and direct labor. Manufacturers' disappointment with FMS has led to a growing preference for the FMC, and partial integration of FMCs into an apparent FMS.

Both FMCs and FMSs employ robots, especially gantry or jointed-arm, as machine tenders. In both cases, planning and simulation of production and production processes are important. Even if the eventual integration of individual FMCs into an FMS is not envisioned, the computer hardware and software for data acquisition and control and production control needs to evaluated for integration capability. This is necessary so that the company does not find itself with virtual islands of automation in which repeated manual entry of data into different computer systems is required.

FMS versus Fixed Automation. Flexible manufacturing systems include computer-controlled machines (typically CNC), automated material handling, and a computer control system. The purpose of an FMS is to manufacture multiple parts and even multiple families of parts in midvolume quantity with minimal lead times (set up, production, and nonproduction times).

Automated equipment is a tool used to meet production schedules to make a profit. The nature of the "bells and whistles, fixed or flexible'' is not as important as the ability of the equipment to make a profit. One of the key steps in achieving profit-making production is manufacturing quality parts with zero defects the first time. Automation should not occur before the relevant manufacturing processes are understood and under control.

Sensors and Transducers. Discrete switches, digital sensors, and analog sensors or transducers are significant to robotics, FMS, and automation in general. Discrete sensors provide the lowest level of information (binary on/off). They are used to indicate if a tool is broken; if a part is available and in the appropriate location; or if a fluid pressure, flow, or level is within the specified ranges. Digital sensors, such as encoders, provide positioning data with incremental or absolute positioning on the rotary or linear encoder. Two-phase output from quadrature encoders can indicate direction of rotation or movement, as well as the incremental amount of encoder steps that were moved. Absolute encoders are similar except they indicate the actual position on the rotation, not just the net change. Analog sensors or transducers provide the capability to change a physical phenomenon to an electrical signal. Sensors or transducers are available to provide a continuously variable electrical output for most parameters within the different forms of energy.

Robotic arms and end-effectors have a wide range of parameters that may need to be sensed depending on the application. Robot arm velocity is often indicated by the electromotive force (EMF) output from tachometer generators incorporated in the arm drive motors, while the position of the arm is determined by either absolute or incremental encoders.

Tactile sensing of end-of-arm tooling has not been very successful in the production environment while force and torque sensing are in use. Tactile sensing is complex and includes detecting touch, force, movement, shape, and orientation. This requires an array of analog sensors that determine force. The shape, orientation, and slip issues are determined by the appropriate CPU with input from the 2-D array of sensors.

Force sensing is still done primarily with (resistive) strain gages. Arrays of strain gages on or near the surface of the grasping tool have not proven appropriate for production. Force sensing on an end-effector has proven effective. Force and torque sensing using mounting bolts in bolt circles can determine forces in three dimensions and moments in three dimensions. Common among the applications that require force and torque sensing is robotic grinding and deburring since the robot arm generally needs to accommodate the wear and change in diameter of the grinding/deburring medium. This wear rate is variable (especially with wheel dressing) and the robot needs to adjust the arm until the appropriate forces and torques are being developed for the part material removal.

Criteria for evaluating or selecting sensors and transducers include the phenomenon being sensed, the range of the phenomenon, the resolution of the sensor, the accuracy, the repeatability, the linearity of the output/input ratio, operating life time or mean time between failure (MTBF), the response time for the sensor, hysteresis, output signal, output due to extraneous variables, and the environmental constraints for the sensor. The signal conditioning from the sensor or transducer will affect the above criteria as much or more than the sensor or transducer itself. The ability of the signal conditioner to correct for sensor error, for output variation due to extraneous variables, and other factors is likewise important. The sensor or transducer and the signal conditioner should always be viewed as a system and not as isolated units. When the sensors are integrated with robots or CNC tools, the complete system response time becomes the issue.

21.10 Boolean Logic

Boolean logic is used in computer programs, spreadsheets, computer numerical control (CNC), programmable logic controller programs, and robotic programs. The classical Boolean operators (Figure 21-7) are OR [+], AND [*], and NOT[⁻]. OR logic is typically depicted schematically in parallel branches in relay ladder logic. In Figure 21-7, if Input 1 OR Input 2 is true, then Output 1 (cella) is true, otherwise Output 1 is false (cellb).

The logical conditions of equal (=), greater than ($>$), or less than ($<$), Figure 21-8, are used in arithmetic and algebraic operations in computer programs, spreadsheets, computer numerical control, programmable logic controls, and robotic programs. These operations are used commonly in conditional branching.

Understanding and applying these operators accurately is fundamental to conditional branching in conventional computer programs, spreadsheets, computer numerical control programs, programmable logic control programs, and robot programs. Expression of a logical operation in robot programs is still a proprietary function.

Boolean Algebra	Computer Program	Spreadsheet
OR		
A +B = C	C = A OR B	@IF(cellx#ORcelly,cellb)
AND		
A * B = C	C = A AND B	@IF(cellx#AND#celly,cella,cellb)
NOT		
A = C	C = NOT A	@IF(NOT#cellx =celly,cella,cellb)

Computer Numerical Control	Relay Ladder Logic
OR	▼ ▼
D1 = (Cond1 **OR** Cond2)	

```
                                    |    Input A           Output  C  |
                                    |-----] [-------|-------------( )----|
                                    ||             |                    |
                                    ||   Input B   |                    |
                                    |-----] [-------|                    |
                                    |                                   |
```

| **AND** | ▼   ▼ |
| D1 = (Cond1 **AND** Cond2) | |

```
                                    |    Input A   Input B   Output C  |
                                    |-----] [-------] [------( )-------|
                                    |                                  |
```

| **NOT** | |
| Not used | |

```
                                    |    Input A           Output C    |
                                    |-----]/[----------------( )-------|
                                    |                                  |
```

Figure 21-7. Boolean Algebra Operators

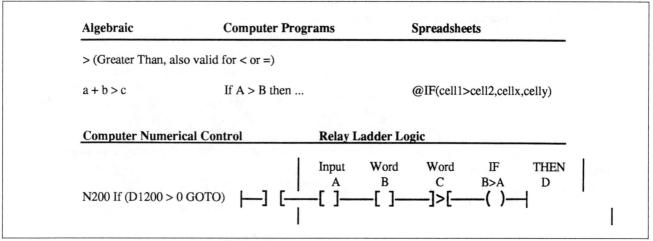

Figure 21-8. Use of Logical Conditions in Various Applications

Practice Problems

21.1 Which of the criteria for manufacturing automation does computer simulation of product performance satisfy?

    a. Quality of work life
    b. Product quality
    c. Product development time
    d. Production efficiency

21.2 Of the three-dimensional CAD modeling systems, which provides the designer the greatest potential for understanding the inside and outside of the model?

    a. Wireframe
    b. Solid
    c. Surface

21.3 What is the function of network protocols?

21.4 The principal shortfall of open-loop CNC motion control is

    a. It is a proprietary interface
    b. It provides no feedback
    c. Its commands are time-based
    d. It requires mainframe-level computing power

21.5 In what way is the "right-hand rule" helpful to the CNC specialist?

21.6 Bar codes, radio frequency data transmission, magnetic stripes, and voice recognition are types of

    a. Optical character recognition systems
    b. Automatic identification technology
    c. Image interpretation systems
    d. Product analysis methodology

Appendix A

Metrics and the SI System

METRICS AND THE SI SYSTEM

Introduction

The metric system is unique in that every industrialized nation stands to gain appreciable economic benefits through a universal system of mathematical logic. Commerce and industry in the United States is planning to make a transition into a specific type of the metric system commonly referred to as International System of Units (SI), the system established in 1960 by the General Conference of Weights and Measures.

The International System of Units (SI) was first released by the Conference Generale des Poids et Mesures (CGPM) in 1960 under the French title, Le Systeme International d'Unites. It has since been referred to in all languages as SI. The United States and approximately 35 other countries have participated in that international conference.

As with any of the systems available throughout the world, the metric system deals in base and derived units for length, mass, time, force, energy, and so forth. The following has been developed to reflect units of measurements and multiples or divisions of those base units for all of the common forms of measurements in the metric SI system.

The SI, like the traditional metric system, is based on decimal arithmetic. Each physical quantity consists of units of different sizes which are formed or created by multiplying or dividing a single base value by powers of 10. Thus, changes can be made very simply by adding zeros or moving decimal points.

Quantity	Unit	Symbol	Base of definition
Length . . . . meter		m	Distance in which light travels in a vacuum in 1/299,792,458 second
Time . . . . second		s	Cycles of radiation of cesium
Mass . . . . kilogram		kg	Platinum-iridium cylinder prototype
Temperature . . . kelvin		K	Absolute zero and water triple point
Electric current . . . . . ampere		A	Force between two conducting wires
Luminous intensity . . . . candela		cd	Intensity of an area of platinum
Quantity of substance . . . . mole		mol	Amount of atoms in carbon 12

Figure A-1. *This figure represents the seven basic SI units established by the General Conference on Weights and Measures (CGPM). (Courtesy of the Tool And Manufacturing Engineers Handbook, Third Edition.)*

THE BASE UNITS

There are seven base units which represent the quantities of length, time, mass, thermodynamic temperature, electric current, luminous intensity, and amount of substance. These units are expressed as the meter (m), the second (s), the kilogram (kg), the kelvin (K), the ampere (A), the candela (cd), and the mole (mol). The table in Figure A-1 represents the seven base SI units established by the General Conference on Weights and Measures (CGPM).

Length. The base unit for length in the SI system is the meter. The millimeter (1/1000 of a meter) is used as a standard unit of length for most drafting and machining practices.

Time. The second is the base unit for time in the SI system. However, the minute, hour and day are acceptable non-SI units. These units are multiples of the second and are used for convenience and familiarity. Figure A-2 shows the relationship of these units.

Mass. The SI unit for mass is the kilogram. The term mass reflects the amount of matter an object contains. The kilogram represents only mass. Weight and force are represented by other SI units.

In addition to objects having mass, liquids also can have a mass equivalency. In the metric system there is a direct relationship between volume and mass of water. It has been established that one decimeter cubed (dm^3) of water has a mass of one kilogram and that one centimeter cubed (cm^3) of water has a mass of one gram. The liter (L), an acceptable non-SI term is equal to one decimeter cubed (dm^3). It should be noted here that the symbol for liter has created some confusion. Formerly, the lower case "el" (l) was the only authorized symbol for liter. In 1979, however, the CGPM, while continuing to sanction the lower case "el" (l), also authorized the capital L as an alternative symbol. In the United States, the capital L now is the officially recommended symbol for liter.

Units	Symbol		Equals
1 minute	min	=	60 seconds
1 hour	h	=	60 minutes or 3600 seconds
1 day	d	=	24 hours or 1440 minutes or 86 400 seconds

Figure A-2 *This figure shows the relationship of the various units of time.*

Temperature. The SI unit for temperature is the kelvin (K). The kelvin scale is developed from the Celsius scale (°C). In the Celsius scale, 0°C equals the freezing point of pure water and 100°C equals the boiling point of pure water at controlled atmospheric pressures.

Zero on the kelvin scale (0 K) represents absolute zero, the temperature at which there is absence of heat. Absolute zero on the Celsius scale is -273.16°C. Therefore, since both scales have equal units, 0°C equals 273.16K. [Note, when writing temperatures in kelvin, the degree symbol (°) is not used.]

Since there is a correlation between kelvin and Celsius, Celsius is preferred and is a non-SI acceptable unit.

Electric Current. The base unit for electric current in the SI, is the ampere (A). The ampere is defined as the amount of current between two straight parallel wires placed one meter apart that results in a specific force between these two wires.

Luminous Intensity. The SI unit for luminous intensity is the candela (cd). The candela is defined as the luminous intensity of an area of a black body at the temperature of freezing platinum under a specific pressure.

Amount of Substance. The mole is the base SI unit for amount of substance. The mole is defined as the amount of substance of a system containing the same amount of elementary entities as there are atoms in carbon 12 having a mass of 0.012 kilogram.

SI PREFIXES

One of the most beneficial aspects of the SI system is the ability to multiply the base unit by powers of 10 to express multiples or sub-multiples of the base unit. These units then are expressed by adding a prefix to the name of the base unit. For example, one thousandth of a meter is expressed as a millimeter, and one thousand meters is expressed as a kilometer. Figure A-3 represents the expression of metric values as noted by the multiplication factors, the prefix names, and the SI symbols.

SI Symbol	Prefix	Multiplication Factors
T	tera	10^{12} = 1 000 000 000 000
G	giga	10^{9} = 1 000 000 000
M	mega	10^{6} = 1 000 000
k	kilo	10^{3} = 1 000
h	hecto*	10^{2} = 100
da	deka*	10^{1} = 10
d	deci*	10^{-1} = 0.1
c	centi*	10^{-2} = 0.01
m	milli	10^{-3} = 0.001
μ	micro	10^{-6} = 0.000 001
n	nano	10^{-9} = 0.000 000 001
p	pico	10^{-12} = 0.000 000 000 001
f	femto	10^{-15} = 0.000 000 000 000 001
a	atto	10^{-18} = 0.000 000 000 000 000 001

* To be avoided when possible

Figure A-3 *This figure represents the expression of metric values as noted by the multiplication factors, the prefix names and the SI symbols.*

THE DERIVED AND SUPPLEMENTARY UNITS

When measuring such quantities as density, pressure, speed, acceleration, and area, it becomes necessary to use combinations of the seven base SI units. These combinations are called *derived units.*

Supplementary units are units which have not been specifically classified as base or derived units. At present, there are only two supplementary units and they both refer to angular measurement. These units are the *radian* and the *steradian.* The more familiar form of angular measurement using degrees did not become an SI supplementary unit. However, the degree (°), the minute ('), and the second (") are acceptable non-SI units. These units and their equivalency are shown in Figure A-4.

Figure A-5 lists the derived units with their appropriate symbols and formulas. Several of the derived units have special names. Figure A-6 is a list of these units and a brief description of each.

Units	Symbol		Equals
1 degree	°	=	$(\pi/180)$rad
1 minute	'	=	1/60°
1 second	"	=	1/60'

Figure A-4 *This figure shows supplementary units and their equivalency.*

Quantity	Unit	Symbol
length	meter	m
mass	kilogram	kg
time	second	s
electric current	ampere	A
thermodynamic temperature	kelvin	K
amount of substance	mole	mol
luminous intensity	candela	cd
plane angle	radian	rad
solid angle	steradian	sr

Quantity	Unit	Symbol	Formula
frequency (of a periodic phenomenon)	hertz	Hz	1/s
force	newton	N	$kg \cdot m/s^{-2}$
pressure, stress	pascal	Pa	N/m^{2}
energy, work, quantity of heat	joule	J	$N \cdot m$
power, radiant flux	watt	W	J/s
quantity of electricity, electric charge	coulomb	C	$A \cdot s$
electric potential, potential difference, electromotive force	volt	V	W/A
electric capacitance	farad	F	C/V
electric resistance	ohm	Ω	V/A
electric conductance	siemens	S	A/V
magnetic flux	weber	Wb	$V \cdot s$
magnetic flux density	tesla	T	Wb/m^{2}
inductance	henry	H	Wb/A
Celsius temperature	degree Celsius[A]	°C	K
luminous flux	lumen	lm	$cd \cdot sr$
illuminance	lux	lx	lm/m^{2}
activity (of a radionuclide)	becquerel	Bq	1/s
absorbed dose[B]	gray	Gy	J/kg
dose equivalent	sievert	Sv	J/kg

[A] Inclusion in the table of derived SI units with special names approved by the CIPM in 1976.

[B] Related quantities using the same unit are: specific energy imparted, kerma, and absorbed dose index.

Figure A-5 *This figure lists SI base, supplementary, and derived units with special names, along with their symbols and formulas. (Adapted from National Bureau of Standards (NBS) Special Publication 330, "The International System of Units (SI)," 1981 edition.)*

DERIVED UNITS

Coulomb	The quantity of electricity transported in one second by a current of one ampere.
Farad	The electric capacitance of a capacitor between the plates of which there appears a difference of potential of one volt when it is charged by a quantity of electricity equal to one coulomb.
Henry	The electric inductance of a closed circuit in which an electromotive force of one volt is produced when the electric current in the circuit varies uniformly at a rate of one ampere per second.
Hertz	A frequency of one cycle per second.
Joule	The work done (i.e., energy expended) when the point of application of a force of one newton is displaced a distance of one meter in the direction of the force.
Lumen	The luminous flux emitted in a solid angle of one steradian by a point source having a uniform intensity of one candela.
Lux	The illuminance produced by a luminous flux of one lumen uniformly distributed over a surface of one square meter.
Newton	That force which, when applied to a body having a mass of one kilogram, gives it an acceleration of one meter per second squared.
Ohm	The electric resistance between two points of a conductor when a constant difference of potential of one volt, applied between these two points, produces in this conductor a current of one ampere, this conductor not being the source of any electromotive force.
Pascal	The pressure or stress of one newton per square meter.
Siemens	The electrical conductance of a conductor in which a current of one ampere is produced by an electric potential difference of one volt.
Tesla	The magnetic flux density given by a magnetic flux of one weber per square meter.
Volt	The volt (unit of electric potential difference and electromotive force) is the difference of electric potential between two points of a conductor carrying a constant current of one ampere, when the power dissipated between these points is equal to one watt.
Watt	The power which gives rise to the production of energy at the rate of one joule per second.
Weber	The magnetic flux which, linking a circuit of one turn, produces in it an electromotive force of one volt as it is reduced to zero at a uniform rate in one second.

The above listed material is from the SI Metric Reference Manual by Lockheed Missiles & Space Company, Inc., Sunnyvale, CA. August 1981 edition.

Figure A-6 *This figure presents a list of derived units and a brief description of each. (The meter spelling is used, in accord with SME editorial practice.)*

Conversions

Quite often, it is necessary to convert from U.S. Customary units (inch, pound, etc.) to SI Metric units. Eventually, the United States will be an SI Metric country and the need for conversions will diminish. Before that time arrives, it will be necessary to provide a means of conversion from one system to the other. Figure A-7 shows the values used when converting from U.S. Customary to SI Metric. These conversion factors are listed by category. Simply locate the U.S. Customary unit in the left column. Multiply this by the conversion factor in the right column. The SI units listed in the middle column are how the new value is expressed.

Notation. Conversion factors are presented for ready adaptation to computer readout and electronic data transmission. The factors are written as a number greater than one and less than ten with six or less decimal places. This number is followed by the letter E (for exponent), a plus or minus symbol, and two digits which indicate the power of 10 by which the number must be multiplied to obtain the correct value. For example:

$$3.523\,907\ E - 02 \text{ is } 3.523\,907 \times 10^{-2}$$

or

$$0.035\,239\,07$$

Similarly:

$$3.386\,389\ E + 03 \text{ is } 3.386\,389 \times 10^{3}$$

or

$$3\,386.389$$

Accuracy. An asterisk (*) after the sixth decimal place indicates that the conversion factor is exact and that all subsequent digits are zero. All other conversion factors have been rounded.

Editor's Note Regarding Spelling: The meter and liter spellings are used generally throughout this text. While recognizing that, in the United States, both the meter and metre and the liter and litre spellings are used, the Society of Manufacturing Engineers (SME), in its editorial practice, is consistent in use of the meter and liter spellings. This usage is in conformance with the highest level of USA SI documentation: The Metric Conversion Act of 1975 (Public Law 94-168)...The Federal Register "Notice", October 26, 1977, "The Metric System of Measurement"...and NBS Special Publication 330, "The International System of Units (SI)", 1981 edition—which is the official USA English-language translation of "Le Systeme International d'Unites (SI)," published in French by the International Bureau of Weights and Measures (BIPM). The meter and liter spellings are used by the National Bureau of Standards, The American National Standards Institute, and the American National Metric Council; and also were used by the United States Metric Board. At the American National Standard level of documentation, different spellings are used in two co-equal authoritative documents "Standard for Metric Practice" (ANSI Z210.1) that otherwise are identical: ASTM E380 (metre and litre); and IEEE Std. 268 (meter and liter).

CLASSIFIED LIST OF UNITS

To convert from	to	Multiply by

ACCELERATION

ft/s²	metre per second squared (m/s²)	3.048 000*E−01
free fall, standard (g)	metre per second squared (m/s²)	9.806 650*E+00
gal	metre per second squared (m/s²)	1.000 000*E−02
in/s²	metre per second squared (m/s²)	2.540 000*E−02

ANGLE

degree	radian (rad)	1.745 329 E−02
minute	radian (rad)	2.908 882 E−04
second	radian (rad)	4.848 137 E−06
grad	degree (angular)	9.000 000*E−01
grad	radian (rad)	1.570 796 E−02

AREA

acre	square metre (m²)	4.046 873 E+03
are	square metre (m²)	1.000 000*E+02
barn	square metre (m²)	1.000 000*E−28
circular mil	square metre (m²)	5.067 075 E−10
darcy	square metre (m²)	9.869 233 E−13
ft²	square metre (m²)	9.290 304*E−02
hectare	square metre (m²)	1.000 000*E+04
in²	square metre (m²)	6.451 600*E−04
mi² (international)	square metre (m²)	2.589 988 E+06
mi² (U.S. statute)	square metre (m²)	2.589 998 E+06
yd²	square metre (m²)	8.361 274 E−01

BENDING MOMENT OR TORQUE

dyne·cm	newton metre (N·m)	1.000 000*E−07
kgf·m	newton metre (N·m)	9.806 650*E+00
ozf·in	newton metre (N·m)	7.061 552 E−03
lbf·in	newton metre (N·m)	1.129 848 E−01
lbf·ft	newton metre (N·m)	1.355 818 E+00

BENDING MOMENT OR TORQUE PER UNIT LENGTH

lbf·ft/in	newton metre per metre (N·m/m)	5.337 866 E+01
lbf·in/in	newton metre per metre (N·m/m)	4.448 222 E+00

CAPACITY (See VOLUME)

DENSITY (See MASS PER UNIT VOLUME)

ELECTRICITY AND MAGNETISM

abampere	ampere (A)	1.000 000*E+01
abcoulomb	coulomb (C)	1.000 000*E+01
abfarad	farad (F)	1.000 000*E+09
abhenry	henry (H)	1.000 000*E−09
abmho	siemens (S)	1.000 000*E+09
abohm	ohm (Ω)	1.000 000*E−09

NOTE: ESU means electrostatic cgs unit. EMU means electromagnetic cgs unit.

Figure A-7 *This chart shows the conversion values used to convert from U.S. Customary to SI Metric. (Courtesy of the American Society For Testing And Materials. Reprinted from Standard E380-82.) Editor's Note: The ASTM spelling practice (metre and litre) is retained in this tabulation of conversion factors.*

To convert from	to	Multiply by
abvolt	volt (V)	1.000 000*E−08
ampere hour	coulomb (C)	3.600 000*E+03
EMU of capacitance	farad (F)	1.000 000*E+09
EMU of current	ampere (A)	1.000 000*E+01
EMU of electric potential	volt (V)	1.000 000*E−08
EMU of inductance	henry (H)	1.000 000*E−09
EMU of resistance	ohm (Ω)	1.000 000*E−09
ESU of capacitance	farad (F)	1.112 650 E−12
ESU of current	ampere (A)	3.335 6 E−10
ESU of electric potential	volt (V)	2.997 9 E+02
ESU of inductance	henry (H)	8.987 554 E+11
ESU of resistance	ohm (Ω)	8.987 554 E+11
faraday (based on carbon-12)	coulomb (C)	9.648 70 E+04
faraday (chemical)	coulomb (C)	9.649 57 E+04
faraday (physical)	coulomb (C)	9.652 19 E+04
gamma	tesla (T)	1.000 000*E−09
gauss	tesla (T)	1.000 000*E−04
gilbert	ampere (A)	7.957 747 E−01
maxwell	weber (Wb)	1.000 000*E−08
mho	siemens (S)	1.000 000*E+00
oersted	ampere per metre (A/m)	7.957 747 E+01
ohm centimetre	ohm metre (Ω·m)	1.000 000*E−02
ohm circular-mil per foot	ohm metre (Ω·m)	1.662 426 E−09
statampere	ampere (A)	3.335 640 E−10
statcoulomb	coulomb (C)	3.335 640 E−10
statfarad	farad (F)	1.112 650 E−12
stathenry	henry (H)	8.987 554 E+11
statmho	siemens (S)	1.112 650 E−12
statohm	ohm (Ω)	8.987 554 E+11
statvolt	volt (V)	2.997 925 E+02
unit pole	weber (Wb)	1.256 637 E−07

ENERGY (Includes WORK)

To convert from	to	Multiply by
British thermal unit (International Table)	joule (J)	1.055 056 E+03
British thermal unit (mean)	joule (J)	1.055 87 E+03
British thermal unit (thermochemical)	joule (J)	1.054 350 E+03
British thermal unit (39°F)	joule (J)	1.059 67 E+03
British thermal unit (59°F)	joule (J)	1.054 80 E+03
British thermal unit (60°F)	joule (J)	1.054 68 E+03
calorie (International Table)	joule (J)	4.186 800*E+00
calorie (mean)	joule (J)	4.190 02 E+00
calorie (thermochemical)	joule (J)	4.184 000*E+00
calorie (15°C)	joule (J)	4.185 80 E+00
calorie (20°C)	joule (J)	4.181 90 E+00
calorie (kilogram, International Table)	joule (J)	4.186 800*E+03
calorie (kilogram, mean)	joule (J)	4.190 02 E+03
calorie (kilogram, thermochemical)	joule (J)	4.184 000*E+03
electronvolt	joule (J)	1.602 19 E−19
erg	joule (J)	1.000 000*E−07
ft·lbf	joule (J)	1.355 818 E+00
ft-poundal	joule (J)	4.214 011 E−02
kilocalorie (International Table)	joule (J)	4.186 800*E+03
kilocalorie (mean)	joule (J)	4.190 02 E+03
kilocalorie (thermochemical)	joule (J)	4.184 000*E+03
kW·h	joule (J)	3.600 000*E+06
therm (European Community)	joule (J)	1.055 06 E+08
therm (U.S.)	joule (J)	1.054 804*E+08
ton (nuclear equivalent of TNT)	joule (J)	4.184 E+09
W·h	joule (J)	3.600 000*E+03
W·s	joule (J)	1.000 000*E+00

Figure A-7 continued

To convert from	to	Multiply by

ENERGY PER UNIT AREA TIME

To convert from	to	Multiply by
Btu (International Table)/(ft^2·s)	watt per square metre (W/m^2)	1.135 653 E+04
Btu (International Table)/(ft^2·h)	watt per square metre (W/m^2)	3.154 591 E+00
Btu (thermochemical)/(ft^2·s)	watt per square metre (W/m^2)	1.134 893 E+04
Btu (thermochemical)/(ft^2·min)	watt per square metre (W/m^2)	1.891 489 E+02
Btu (thermochemical)/(ft^2·h)	watt per square metre (W/m^2)	3.152 481 E+00
Btu (thermochemical)/(in^2·s)	watt per square metre (W/m^2)	1.634 246 E+06
cal (thermochemical)/(cm^2·min)	watt per square metre (W/m^2)	6.973 333 E+02
cal (thermochemical)/(cm^2·s)	watt per square metre (W/m^2)	4.184 000*E+04
erg/(cm^2·s)	watt per square metre (W/m^2)	1.000 000*E−03
W/cm^2	watt per square metre (W/m^2)	1.000 000*E+04
W/in^2	watt per square metre (W/m^2)	1.550 003 E+03

FLOW (See MASS PER UNIT TIME or VOLUME PER UNIT TIME)

FORCE

To convert from	to	Multiply by
dyne	newton (N)	1.000 000*E−05
kilogram-force	newton (N)	9.806 650*E+00
kilopond (kp)	newton (N)	9.806 650*E+00
kip (1000 lbf)	newton (N)	4.448 222 E+03
ounce-force	newton (N)	2.780 139 E−01
pound-force (lbf)	newton (N)	4.448 222 E+00
lbf/lb (thrust/weight [mass] ratio)	newton per kilogram (N/kg)	9.806 650 E+00
poundal	newton (N)	1.382 550 E−01
ton-force (2000 lbf)	newton (N)	8.896 444 E+03

FORCE PER UNIT AREA (See PRESSURE)

FORCE PER UNIT LENGTH

To convert from	to	Multiply by
lbf/ft	newton per metre (N/m)	1.459 390 E+01
lbf/in	newton per metre (N/m)	1.751 268 E+02

HEAT

To convert from	to	Multiply by
Btu (International Table)·ft/(h·ft^2·°F) (thermal conductivity)	watt per metre kelvin [(W/(m·K)]	1.730 735 E+00
Btu (thermochemical)·ft/(h·ft^2·°F) (thermal conductivity)	watt per metre kelvin [(W/(m·K)]	1.729 577 E+00
Btu (International Table)·in/(h·ft^2·°F) (thermal conductivity)	watt per metre kelvin [(W/(m·K)]	1.442 279 E−01
Btu (thermochemical)·in(h·ft^2·°F) (thermal conductivity)	watt per metre kelvin [(W/(m·K)]	1.441 314 E−01
Btu (International Table)·in/(s·ft^2·°F) (thermal conductivity)	watt per metre kelvin [(W/(m·K)]	5.192 204 E+02
Btu (thermochemical)·in/(s·ft^2·°F) (thermal conductivity)	watt per metre kelvin [(W/(m·K)]	5.188 732 E+02
Btu (International Table)/ft^2	joule per square metre (J/m^2)	1.135 653 E+04
Btu (thermochemical)/ft^2	joule per square metre (J/m^2)	1.134 893 E+04
Btu (International Table)/(h·ft^2·°F) (thermal conductance)	watt per square metre kelvin [(W/(m^2·K)]	5.678 263 E+00
Btu (thermochemical)/(h·ft^2·°F)(thermal conductance)	watt per square metre kelvin [(W/(m^2·K)]	5.674 466 E+00
Btu (International Table)/(s·ft^2·°F)	watt per square metre kelvin [(W/(m^2·K)]	2.044 175 E+04
Btu (thermochemical)/(s·ft^2·°F)	watt per square metre kelvin [(W/(m^2·K)]	2.042 808 E+04
Btu (International Table)/lb	joule per kilogram (J/kg)	2.326 000*E+03
Btu (thermochemical)/lb	joule per kilogram (J/kg)	2.324 444 E+03
Btu (International Table)/(lb·°F) (heat capacity)	joule per kilogram kelvin [(J/(kg·K)]	4.186 800*E+03
Btu (thermochemical)/(lb·°F) (heat capacity)	joule per kilogram kelvin [(J/(kg·K)]	4.184 000*E+03

Figure A-7 continued

To convert from	to	Multiply by
Btu (International Table)/ft^3	joule per cubic metre (J/m^3)	3.725 895 E+04
Btu (thermochemical)/ft^3	joule per cubic metre (J/m^3)	3.723 402 E+04
cal (thermochemical)/(cm·s·°C)	watt per metre kelvin [(W/(m·K)]	4.184 000*E+02
cal (thermochemical)/cm^2	joule per square metre (J/m^2)	4.184 000*E+04
cal (thermochemical)/(cm^2·min)	watt per square metre (W/m^2)	6.973 333 E+02
cal (thermochemical)/(cm^2·s)	watt per square metre (W/m^2)	4.184 000*E+04
cal (International Table)/g	joule per kilogram (J/kg)	4.186 800*E+03
cal (thermochemical)/g	joule per kilogram (J/kg)	4.184 000*E+03
cal (International Table)/(g·°C)	joule per kilogram kelvin [(J/kg·K)]	4.186 800*E+03
cal (thermochemical)/(g·°C)	joule per kilogram kelvin [(J/(kg·K)]	4.184 000*E+03
cal (thermochemical)/min	watt (W)	6.973 333 E−02
cal (thermochemical)/s	watt (W)	4.184 000*E+00
clo	kelvin square metre per watt (K·m^2/W)	2.003 712 E−01
°F·h·ft^2/Btu (International Table) (thermal resistance)	kelvin square metre per watt (K·m^2/W)	1.761 102 E−01
°F·h·ft^2/Btu (thermochemical) (thermal resistance)	kelvin square metre per watt (K·m^2/W)	1.762 280 E−01
°F·h·ft^2/[Btu (International Table)·in] (thermal resistivity)	kelvin metre per watt (K·m/W)	6.933 471 E+00
°F·h·ft^2/[Btu (thermochemical)·in] (thermal resistivity)	kelvin metre per watt (K·m/W)	6.938 113 E+00
ft^2/h (thermal diffusivity)	square metre per second (m^2/s)	2.580 640*E−05

LENGTH

angstrom	metre (m)	1.000 000*E−10
astronomical unit	metre (m)	1.495 979 E+11
chain	metre (m)	2.011 684 E+01
fathom	metre (m)	1.828 804 E+00
fermi (femtometre)	metre (m)	1.000 000*E−15
foot	metre (m)	3.048 000*E−01
foot (U.S. survey)	metre (m)	3.048 006 E−01
inch	metre (m)	2.540 000*E−02
light year	metre (m)	9.460 55 E+15
microinch	metre (m)	2.540 000*E−08
micron	metre (m)	1.000 000*E−06
mil	metre (m)	2.540 000*E−05
mile (international nautical)	metre (m)	1.852 000*E+03
mile (U.S. nautical)	metre (m)	1.852 000*E+03
mile (international)	metre (m)	1.609 344*E+03
mile (U.S. statute)	metre (m)	1.609 347 E+03
parsec	metre (m)	3.085 678 E+16
pica (printer's)	metre (m)	4.217 518 E−03
point (printer's)	metre (m)	3.514 598*E−04
rod	metre (m)	5.029 210 E+00
yard	metre (m)	9.144 000*E−01

LIGHT

cd/in^2	candela per square metre (cd/m^2)	1.550 003 E+03
footcandle	lux (lx)	1.076 391 E+01
footlambert	candela per square metre (cd/m^2)	3.426 259 E+00
lambert	candela per square metre (cd/m^2)	3.183 099 E+03
lm/ft^2	lumen per square metre (lm/m^2)	1.076 391 E+01

MASS

carat (metric)	kilogram (kg)	2.000 000*E−04
grain	kilogram (kg)	6.479 891*E−05
gram	kilogram (kg)	1.000 000*E−03
hundredweight (long)	kilogram (kg)	5.080 235 E+01
hundredweight (short)	kilogram (kg)	4.535 924 E+01

Figure A-7 continued

To convert from	to	Multiply by
kgf·s²/m (mass)	kilogram (kg)	9.806 650*E+00
ounce (avoirdupois)	kilogram (kg)	2.834 952 E−02
ounce (troy or apothecary)	kilogram (kg)	3.110 348 E−02
pennyweight	kilogram (kg)	1.555 174 E−03
pound (lb avoirdupois)	kilogram (kg)	4.535 924 E−01
pound (troy or apothecary)	kilogram (kg)	3.732 417 E−01
slug	kilogram (kg)	1.459 390 E+01
ton (assay)	kilogram (kg)	2.916 667 E−02
ton (long, 2240 lb)	kilogram (kg)	1.016 047 E+03
ton (metric)	kilogram (kg)	1.000 000*E+03
ton (short, 2000 lb)	kilogram (kg)	9.071 847 E+02
tonne	kilogram (kg)	1.000 000*E+03

MASS PER UNIT AREA

oz/ft²	kilogram per square metre (kg/m²)	3.051 517 E−01
oz/yd²	kilogram per square metre (kg/m²)	3.390 575 E−02
lb/ft²	kilogram per square metre (kg/m²)	4.882 428 E+00

MASS PER UNIT CAPACITY (See MASS PER UNIT VOLUME)

MASS PER UNIT LENGTH

denier	kilogram per metre (kg/m)	1.111 111 E−07
lb/ft	kilogram per metre (kg/m)	1.488 164 E+00
lb/in	kilogram per metre (kg/m)	1.785 797 E+01
tex	kilogram per metre (kg/m)	1.000 000*E−06

MASS PER UNIT TIME (Includes FLOW)

perm (0°C)	kilogram per pascal second square metre [kg/(Pa·s·m²)]	5.721 35 E−11
perm (23°C)	kilogram per pascal second square metre [kg/(Pa·s·m²)]	5.745 25 E−11
perm·in (0°C)	kilogram per pascal second metre [kg/(Pa·s·m)]	1.453 22 E−12
perm·in (23°C)	kilogram per pascal second metre [kg/(Pa·s·m)]	1.459 29 E−12
lb/h	kilogram per second (kg/s)	1.259 979 E−04
lb/min	kilogram per second (kg/s)	7.559 873 E−03
lb/s	kilogram per second (kg/s)	4.535 924 E−01
lb/(hp·h) (SFC, specific fuel consumption)	kilogram per joule (kg/J)	1.689 659 E−07
ton (short)/h	kilogram per second (kg/s)	2.519 958 E−01

MASS PER UNIT VOLUME (Includes DENSITY and MASS CAPACITY)

grain/gal (U.S. liquid)	kilogram per cubic metre (kg/m³)	1.711 806 E−02
g/cm³	kilogram per cubic metre (kg/m³)	1.000 000*E+03
oz (avoirdupois)/gal (U.K. liquid)	kilogram per cubic metre (kg/m³)	6.236 021 E+00
oz (avoirdupois)/gal (U.S. liquid)	kilogram per cubic metre (kg/m³)	7.489 152 E+00
oz (avoirdupois)/in³	kilogram per cubic metre (kg/m³)	1.729 994 E+03
lb/ft³	kilogram per cubic metre (kg/m³)	1.601 846 E+01
lb/in³	kilogram per cubic metre (kg/m³)	2.767 990 E+04
lb/gal (U.K. liquid)	kilogram per cubic metre (kg/m³)	9.977 633 E+01
lb/gal (U.S. liquid)	kilogram per cubic metre (kg/m³)	1.198 264 E+02
lb/yd³	kilogram per cubic metre (kg/m³)	5.932 764 E−01
slug/ft³	kilogram per cubic metre (kg/m³)	5.153 788 E+02
ton (long)/yd³	kilogram per cubic metre (kg/m³)	1.328 939 E+03
ton (short)/yd³	kilogram per cubic metre (kg/m³)	1.186 553 E+03

POWER

Btu (International Table)/h	watt (W)	2.930 711 E−01

Figure A-7 continued

To convert from	to	Multiply by
Btu (International Table)/s	watt (W)	1.055 056 E+03
Btu (thermochemical)/h	watt (W)	2.928 751 E−01
Btu (thermochemical)/min	watt (W)	1.757 250 E+01
Btu (thermochemical)/s	watt (W)	1.054 350 E+03
cal (thermochemical)/min	watt (W)	6.973 333 E−02
cal (thermochemical)/s	watt (W)	4.184 000*E+00
erg/s	watt (W)	1.000 000*E−07
ft·lbf/h	watt (W)	3.766 161 E−04
ft·lbf/min	watt (W)	2.259 697 E−02
ft·lbf/s	watt (W)	1.355 818 E+00
horsepower (550 ft·lbf/s)	watt (W)	7.456 999 E+02
horsepower (boiler)	watt (W)	9.809 50 E+03
horsepower (electric)	watt (W)	7.460 000*E+02
horsepower (metric)	watt (W)	7.354 99 E+02
horsepower (water)	watt (W)	7.460 43 E+02
horsepower (U.K.)	watt (W)	7.457 0 E+02
kilocalorie (thermochemical)/min	watt (W)	6.973 333 E+01
kilocalorie (thermochemical)/s	watt (W)	4.184 000*E+03
ton of refrigeration (= 12 000 Btu/h)	watt (W)	3.517 E+03

PRESSURE OR STRESS (FORCE PER UNIT AREA)

atmosphere, standard	pascal (Pa)	1.013 250*E+05
atmosphere, technical (= 1 kgf/cm²)	pascal (Pa)	9.806 650*E+04
bar	pascal (Pa)	1.000 000*E+05
centimetre of mercury (0°C)	pascal (Pa)	1.333 22 E+03
centimetre of water (4°C)	pascal (Pa)	9.806 38 E+01
dyne/cm²	pascal (Pa)	1.000 000*E−01
foot of water (39.2°F)	pascal (Pa)	2.988 98 E+03
gf/cm²	pascal (Pa)	9.806 650*E+01
inch of mercury (32°F)	pascal (Pa)	3.386 38 E+03
inch of mercury (60°F)	pascal (Pa)	3.376 85 E+03
inch of water (39.2°F)	pascal (Pa)	2.490 82 E+02
inch of water (60°F)	pascal (Pa)	2.488 4 E+02
kgf/cm²	pascal (Pa)	9.806 650*E+04
kgf/m²	pascal (Pa)	9.806 650*E+00
kgf/mm²	pascal (Pa)	9.806 650*E+06
kip/in² (ksi)	pascal (Pa)	6.894 757 E+06
millibar	pascal (Pa)	1.000 000*E+02
millimetre of mercury (0°C)	pascal (Pa)	1.333 22 E+02
poundal/ft²	pascal (Pa)	1.488 164 E+00
lbf/ft²	pascal (Pa)	4.788 026 E+01
lbf/in² (psi)	pascal (Pa)	6.894 757 E+03
psi	pascal (Pa)	6.894 757 E+03
torr (mmHg, 0°C)	pascal (Pa)	1.333 22 E+02

RADIATION UNITS

curie	becquerel (Bq)	3.700 000*E+10
rad	gray (Gy)	1.000 000*E−02
rem	sievert (Sv)	1.000 000*E−02
roentgen	coulomb per kilogram (C/kg)	2.58 E−04

SPEED (See VELOCITY)

STRESS (See PRESSURE)

TEMPERATURE

degree Celsius	kelvin (K)	$T_K = t_{°C} + 273.15$
degree Fahrenheit	degree Celsius (°C)	$t_{°C} = (t_{°F} - 32)/1.8$
degree Fahrenheit	kelvin (K)	$T_K = (t_{°F} + 459.67)/1.8$
degree Rankine	kelvin (K)	$T_K = T_{°R}/1.8$
kelvin	degree Celsius (°C)	$t_{°C} = T_K - 273.15$

Figure A-7 continued

To convert from	to	Multiply by

TIME

To convert from	to	Multiply by
day	second (s)	8.640 000*E+04
day (sidereal)	second (s)	8.616 409 E+04
hour	second (s)	3.600 000*E+03
hour (sidereal)	second (s)	3.590 170 E+03
minute	second (s)	6.000 000*E+01
minute (sidereal)	second (s)	5.983 617 E+01
second (sidereal)	second (s)	9.972 696 E−01
year (365 days)	second (s)	3.153 600*E+07
year (sidereal)	second (s)	3.155 815 E+07
year (tropical)	second (s)	3.155 693 E+07

TORQUE (See BENDING MOMENT)

VELOCITY (Includes SPEED)

To convert from	to	Multiply by
ft/h	metre per second (m/s)	8.466 667 E−05
ft/min	metre per second (m/s)	5.080 000*E−03
ft/s	metre per second (m/s)	3.048 000*E−01
in/s	metre per second (m/s)	2.540 000*E−02
km/h	metre per second (m/s)	2.777 778 E−01
knot (international)	metre per second (m/s)	5.144 444 E−01
mi/h (international)	metre per second (m/s)	4.470 400*E−01
mi/min (international)	metre per second (m/s)	2.682 240*E+01
mi/s (international)	metre per second (m/s)	1.609 344*E+03
mi/h (international)	kilometre per hour (km/h)	1.609 344*E+00

VISCOSITY

To convert from	to	Multiply by
centipoise (dynamic viscosity)	pascal second (Pa·s)	1.000 000*E−03
centistokes (kinematic viscosity)	square metre per second (m²/s)	1.000 000*E−06
ft²/s	square metre per second (m²/s)	9.290 304*E−02
poise	pascal second (Pa·s)	1.000 000*E−01
poundal·s/ft²	pascal second (Pa·s)	1.488 164 E+00
lb/(ft·h)	pascal second (Pa·s)	4.133 789 E−04
lb/(ft·s)	pascal second (Pa·s)	1.488 164 E+00
lbf·s/ft²	pascal second (Pa·s)	4.788 026 E+01
lbf·s/in²	pascal second (Pa·s)	6.894 757 E+03
rhe	1 per pascal second [1/(Pa·s)]	1.000 000*E+01
slug/(ft·s)	pascal second (Pa·s)	4.788 026 E+01
stokes	square metre per second (m²/s)	1.000 000*E−04

VOLUME (Includes CAPACITY)

To convert from	to	Multiply by
acre-foot	cubic metre (m³)	1.233 489 E+03
barrel (oil, 42 gal)	cubic metre (m³)	1.589 873 E−01
board foot	cubic metre (m³)	2.359 737 E−03
bushel (U.S.)	cubic metre (m³)	3.523 907 E−02
cup	cubic metre (m³)	2.365 882 E−04
fluid ounce (U.S.)	cubic metre (m³)	2.957 353 E−05
ft³	cubic metre (m³)	2.831 685 E−02
gallon (Canadian liquid)	cubic metre (m³)	4.546 090 E−03
gallon (U.K. liquid)	cubic metre (m³)	4.546 092 E−03
gallon (U.S. dry)	cubic metre (m³)	4.404 884 E−03
gallon (U.S. liquid)	cubic metre (m³)	3.785 412 E−03
gill (U.K.)	cubic metre (m³)	1.420 654 E−04
gill (U.S.)	cubic metre (m³)	1.182 941 E−04
in³	cubic metre (m³)	1.638 706 E−05
litre	cubic metre (m³)	1.000 000*E−03
ounce (U.K. fluid)	cubic metre (m³)	2.841 307 E−05

Although speedometers may read km/h, the SI unit is m/s.

Figure A-7 continued

To convert from	to	Multiply by
ounce (U.S. fluid)	cubic metre (m^3)	2.957 353 E−05
peck (U.S.)	cubic metre (m^3)	8.809 768 E−03
pint (U.S. dry)	cubic metre (m^3)	5.506 105 E−04
pint (U.S. liquid)	cubic metre (m^3)	4.731 765 E−04
quart (U.S. dry)	cubic metre (m^3)	1.101 221 E−03
quart (U.S. liquid)	cubic metre (m^3)	9.463 529 E−04
stere	cubic metre (m^3)	1.000 000*E+00
tablespoon	cubic metre (m^3)	1.478 676 E−05
teaspoon	cubic metre (m^3)	4.928 922 E−06
ton (register)	cubic metre (m^3)	2.831 685 E+00
yd^3	cubic metre (m^3)	7.645 549 E−01

VOLUME PER UNIT TIME (Includes FLOW)

ft^3/min	cubic metre per second (m^3/s)	4.719 474 E−04
ft^3/s	cubic metre per second (m^3/s)	2.831 685 E−02
gallon (U.S. liquid)/(hp·h)(SFC, specific fuel consumption)	cubic metre per joule (m^3/J)	1.410 089 E−09
in^3/min	cubic metre per second (m^3/s)	2.731 177 E−07
yd^3/min	cubic metre per second (m^3/s)	1.274 258 E−02
gallon (U.S. liquid) per day	cubic metre per second (m^3/s)	4.381 264 E−08
gallon (U.S. liquid) per minute	cubic metre per second (m^3/s)	6.309 020 E−05

WORK (See ENERGY)

Figure A-7 continued

Appendix B

Answers to the Practice Problems

ANSWERS TO THE PRACTICE PROBLEMS

1.1 $x=4$

1.2 $x=-2$

1.3 $x=3y/(1-2y)$

1.4 $x=5$

1.5 $w=9$

1.6 $x=82$

1.7 $y_{1,2}=-3.62,-1.38$

1.8 $g_{1,2}=-3.19,2.1\overline{9}$

1.9 $r_{1,2}=-3.5,0.33\overline{3}$

1.10 $x_{1,2}=1\pm2\sqrt{-2}$

1.11 $(8-x)(8+x)=64-9$, $64-x^2=55$, $x^2=9$, $x=3$

1.12 $2(30+2x)x+2(40)x=296$, $x^2+35x-74=0$, $x=2$ *(only one valid solution)*

1.13 $2((w-4)-4)(w-4)=256$, $w^2-12w+32=128$, $w=17.5$

1.14 $x=3.5$, $y=-1$

1.15 $x=5$, $y=5$

1.16 $x=2$, $y=1$

1.17 $x=1.5$

1.18 $x=8$

1.19 $x=258.2$

1.20 $A=100\pi/4+10^2\pi=125\pi=393$, *paint*$=393/5=78.5\,l$

1.21 $y=-2x+4$

1.22 $y=1.5x-6$

1.23 $x^2/(150/2)^2 + y^2/(75/2)^2=1$

1.24 $(x-5)^2+(y-8)^2=8^2$

1.25 $x^2-y^2=1$

1.26   Parabola

1.27 $\cos^{-1}0.8=36.9°$

1.28 $3^2+4^2=H^2$, $H=5$

1.29 $180°-90°-x+36.9°$, $x=53.1°$

1.30 $29°$, $47°$, $104°$

1.31 $2\tan\theta$

1.32 $1-2\sin^2\theta$

1.33 7.2

1.34   a. $4/52=1/13$

       b. $2/52=1/26$

       c. $1/4$

       d. $1/52$

       e. $12/52=3/13$

1.35   a. $6/36=1/6$

       b. $P(not\,2)=1-P(2)=35/36$

       c. $18/36=1/2$

1.36 $6!=720$

1.37 $(26)(25)(24)(10)(9)(8)=11.232 \times 10^6$

1.38 $(30!)/(27!\,3!)=4060$

1.39 $(52!)/(50!\,2!)=1326$

1.40 $8/52$, $39/52$

1.41 $P(H\;and\;H)=P(H)P(H)=1/4$

1.42   a. $(1/2)(1/2)(1/2)=1/8$

       b. $(1/2)(1/2)(1/2)=1/8$

1.43   a. 75

       b. 70.2

       c. 13.6

1.44 $15x^4+14x$

1.45 $3x^2\sin x+x^3\cos x$

1.46 $(x^2+4)/(3x^2)$

1.47 3

1.48 $x=1$

1.49 15

1.50 24.1

2.1   a, c, f, h, and i are vectors

2.2   a. $11i+14j$

       b. $8i-4j$

       c. 6.7

       d. $0.27i-0.96j$

       e. -57

       f. $73.9°$

       g. $0i+0j-48k$

       h. $0i+0j+1k$

3.1	$\sim 10^{14} Hz$	7.6	141.4 N m CW
3.2	$670 \times 10^{-9} m$	7.7	Slide
3.3	No, $3.41 \times 10^{14} Hz$	7.8	243 lb
3.4	Green	7.9	$\bar{x} = 1.5, \bar{y} = 2.46$
3.5	7.5	7.10	$\bar{x} = -0.84, \bar{y} = 0$
3.6	1.41		
3.7	2.31 cm	8.1	90 m
3.8	15 cm, real, inverted	8.2	10.2 s
3.9	27 cm	8.3	7.23 m
3.10	-13 cm	8.4	47.7 rev
		8.5	6.2 s
4.1	30 dB	8.6	$17.2 \ ft/s^2$
4.2	70 dB, 600 Hz	8.7	10 rad/s
4.3	the density also increases	8.8	87.5 m
4.4	$10^{-3} W/m^2$	8.9	1:1
4.5	66 dB	8.10	5.58 in
4.6	$8 \times 10^{-5} W$	8.11	$6.21 \times 10^6 lb$
4.7	93 Hz	8.12	$7.84 \ rad/s^2$
5.1	a. $146°C$	9.1	4529 psi
	b. $755°R$	9.2	0.0017 mm
	c. $419°K$	9.3	40×10^6 psi
5.2	$9.5 \times 10^{-4} \ in$	9.4	7.9
5.3	$178°F$	9.5	0.1 in
5.4	$71°F$	9.6	10300 lb
5.5	10 kg	9.7	41 in lb
5.6	increases	9.8	0.098 rad
5.7	A larger quantity of heat is discharged into the house than is removed from the food.	9.9	22.2 psi
		9.10	21.9 psi
5.8	Copper	9.11	0.311 in
5.9	1000 cal/s	9.12	160 lb
5.10	an upper bound of 2.97 s	9.13	0.48 in × 0.48 in
5.11	5888 W		
		10.1	obstructions in orderly crystal growth
6.1	24 W	10.2	lower
6.2	0.12 A	10.3	0.008%-1%
6.3	I = 9.1 A, R = 12.1 Ω	10.4	$727°C$
6.4	electrical energy	10.5	avoid martensite formation
6.5	series	10.6	275 BHN
6.6	1 μF	10.7	71000 psi
6.7	7.5 Ω	10.8	$\sim 850°C$
6.8	I = 4.5 A	10.9	lead
6.9	I = 5.15 A	10.10	chromium
6.10	8000 V/m	10.11	gray iron
6.11	Down	10.12	nodular iron
		10.13	magnesium
7.1	$R - Pa/(a+b)$	11.1	$kg \cdot m/s^2$
7.2	100 lbs	11.2	1.44×10^6
7.3	T = 1000 lb	11.3	4.7 psi
7.4	283 lb		
7.5	1000 lb horizonal, 4000 ft lb CCW		

11.4   4.21 kPa
11.5   1312 lb
11.6   3.04 ft^3
11.7   31.3 m/s
11.8   894 kPa
11.9   19 cm
11.10 4.92 m/s

12.4   nominal
12.5   datum
12.6   0.1 in
12.7   hole
12.8   both
12.9   50 - basic size, f - shaft, 6 - IT Grade
12.10 straightness, flatness, circularity, cylindricity
12.11 refer to Section 12 examples
12.12 refer to Section 12 examples
12.13 a. roughness height
       b. waviness height
       c. lay

13.1   1.0002 lower limit, e.g. 1.0003±0.0001

13.2 $\dfrac{0.508}{0.505}$, $0.505\,^{+0.003}_{-0.000}$, $0.506\,^{+0.002}_{-0.001}$

13.3   statistical

13.4   pin $1.000\,^{+0.000}_{-0.001}$, hole $1.001\,^{+0.001}_{-0.000}$

13.5   0.0025 in
13.6   0.0051 in
13.7   0.001 in
13.8   0.007 in

13.9   Go: $2.1498\,^{+0.0000}_{-0.0002}$ No Go: $2.1450\,^{+0.0002}_{-0.0000}$

13.10 Go: $2.991\,^{+0.001}_{-0.000}$ No Go: $3.010\,^{+0.000}_{-0.001}$

14.1   Variables: diameter, weight
       Attributes: Go/No Go, Broken/Intact
14.2   Assignable: Operator Error, Tool Wear
       Natural: Material Variation, Tool Composition
14.3   standard deviation, range
14.4   2.503, 0.001516
14.5   1.35
14.6   77.85, 72.85 9.165, 0
14.7   23.43, 22.13 5.324, 0
14.8   1.863
14.9   0.588

15.1 $527
15.2 $365
15.3 $15,590
15.4 $134,200
15.5 $8,222
15.6   No, the most one should be willing to pay is $2668 per year.
15.7 $10,722
15.8   96 checks

16.1   74.4 in (1890 mm)
16.2   25.2 in (640 mm)
16.3   50 lux
16.4   200 lux
16.5   90 dBA for eight hour exposure
16.6   resonance of internal organs
16.7   CTD, e.g. carpal tunnel syndrome

17.1. Fatigue properties of a material under test and the number of stress cycles before failure.
17.2. Proportion limit, elastic limit, yield point, yield strength, ultimate strength, breaking (rupture) strength, ductility, and modulus of elasticity.
17.3. Ductility, hardness, toughness.
17.4. Ductility—The amount of deformation of material can withstand until failure, as determined by the tensile test.
Hardness—That property of a material that enables it to resist plastic deformation.
Toughness—A material's ability to absorb energy from impact.
17.5. High surface wear resistance.
17.6. d.
17.7. d.
17.8. c.

18.1. (1) Organization of process components and standardization.
       (2) Length of production run.
       (3) Complexity of scheduling procedures.
18.2. a.
18.3. a.
18.4. c.
18.5. c.

19.1. 75
19.2. 11 sec
19.3. 14.39°
19.4. 382 rpm
19.5. 0.385 in.

19.6. 196 fm

19.7. 0.702 hp

19.8. 0.00070 in. per tooth

19.9. 1.0 hp

19.10. 0.455 in.

19.11. 0.05 in.

19.12. 29.45 tons

19.13. 75,398 ft-lb

19.14. 389 lb

19.15. 20.94 radians per second^2

19.16. 18,256 psi

19.17. 4500 joules

20.1 Legally, the company's owners are ultimately responsible for the ethical behavior of everyone in the organization. The company will probably discharge the sales director, head of R & D, and manufacturing manager as well.

20.2 100% function, 0% esteem, 0% waste

20.3 The EOQ is 2236.

20.4 b. 20%

20.5 c.

20.6 d.

20.7 b.

20.8 Predictive maintenance procedures determine when equipment breakdown is likely to occur, while preventive maintenance focuses on designing equipment and procedures that reduce or eliminate breakdowns.

21.1 c.

21.2 b.

21.3 Permit access to a computer network

21.4 b.

21.5 Keeping axes in the correct orientation

21.6 b.

Appendix C

CNC G and M Charts

G-Codes for Fanuc OTC Controller
for CNC Turning Center

G-Code	Function
G00	Positioning (Rapid Traverses)
G01	Linear Interpolation (Feed)
G02	Circular Interpolation CW
G03	Circular Interpolation CCW
G04	Dwell
G10	Data Setting
G20	Inch Data Input
G21	Metric Data Input
G22	Stored Stroke Limit Function On
G23	Stored Stroke Limit Function Off
G25	Spindle Speed Change Detection Off
G26	Spindle Speed Change Detection On
G27	Reference Point Return Check
G28	Return to Reference Point
G30	Second Reference Point Return
G31	Skip Function
G32	Threading
G34	Variable Lead Threading
G36	Automatic Tool Compensation X
G37	Automatic Tool Compensation Z
G40	Tool Nose Radius Compensation Cancel
G41	Tool Nose Radius Compensation Left
G42	Tool Nose Radius Compensation Right
G50	Max. Speed Setting of Spindle
G65	Custom Marco Call
G68	Opposed Tool Post Mirror Image On
G69	Opposed Tool Post Mirror Image Off
G70	Finish Cycle
G71	Rough Cutting Cycle of Outer Diameter
G72	Rough Cutting Cycle End Face
G73	Closed Loop Cutting Cycle
G74	End, Face Cutting Off Cycle
G75	Outer/Inner Diameter Cutting Off Cycle
G76	Threading Cycle
G80	Canned Cycle for Drilling
G83	Canned Cycle for Drilling
G84	Canned Cycle for Drilling
G86	Canned Cycle for Drilling
G87	Canned Cycle for Drilling
G88	Canned Cycle for Drilling
G89	Canned Cycle for Drilling
G90	Outer/Inner Cutting Cycle
G92	Threading Cycle
G94	End Face Cutting Cycle
G96	Constant Surface Speed Control
G97	Constant Surface Speed Control Cancel
G98	Feed per Minute
G99	Feed Per Revolution

M-Codes for Fanuc OTC Controller
for a CNC Turning Center*

M-Code	Function
M00	Program Stop
M01	Program Optional Stop
M02	End of Program
M03	Spindle CW
M04	Spindle CCW
M05	Spindle Stop
M08	Cutting Oil On
M09	Cutting Oil Off
M10	Chuck Clamp
M11	Chuck Unclamp
M12	Quill Out
M13	Quill In
M17	Auto Door Close
M18	Auto Door Open
M19	Orientation On
M20	Orientation Off
M21	Door Interlock Bypass On
M22	Door Interlock Bypass Off
M23	Chamfering On
M24	Chamfering Off
M25	Bar Feeder Extend
M28	Part Catcher Extend
M29	Part Catcher Retract
M30	End of Program–Tape Rewind
M31	Chuck Bypass On
M32	Chuck Bypass Off
M43	…
M44	…
M47	Soft Limit 2 Valid
M48	Soft Limit 3 Valid
M49	Soft Limit 2/3 Unvalid
M51	Error Detect Off
M52	Error Detect On
M53	…
M54	…
M55	…
M56	…
M61	X-Axis Mirror Image Off
M71	X-Axis Mirror Image On
M81	On/Off Momentary
M82	On/Off Momentary
M83	M83On M94Off
M84	M84On M93Off
M85	On-Finish-Off
M86	On-Finish-Off
M87	On-Finish-Off
M88	On-Finish-Off
M89	On-Finish-Off
M97	Parts Counted
M98	Calling of Subprogram
M99	M99 Has Rewind Mainprogram

M-Codes are standardized, but many vary from CNC to CNC machine.

General M-Codes* Used in Milling and Turning

M-Code	Function
M00	Program Stop
M01	Optional Stop
M02	End of Program and Tape Rewind
M03	Spindle Start CW
M04	Spindle Start CCW
M05	Spindle Stop
M06	Tool Change
M08	Coolant On
M09	Coolant Off
M19	Spindle Orient and Stop
M21	Mirror Image X
M22	Mirror Image Y
M23	Mirror Image Off
M30	End of Program and Memory Rewind
M41	Low Range
M42	High Range
M48	Override Cancel Off
M49	Override Cancel On
M98	Go to Subroutine
M99	Return from Subroutine

*M-Codes are standardized but many vary from CNC to CNC machine.

G-Codes for Mitsubishi Meldas MO Controller Vertical Machining Center

G-Code	Function
G00	Position (rapid traverse)
G01	Linear Cutting/linear interpolation
G02	Arc interpolation CW/helical cutting CW/R specification arc CW
G03	Arc interpolation CCW/helical cutting CCW/R specification arc CCW
G04	Dwell
G09	Automatic deceleration
G10	Program tool compensation input (compensation amount setting)
G11	Program tool compensation input (compensation amount transfer)
G12	Arc cutting CW
G13	Arc cutting CCW
G14	Coordinate reading function
G15	Print output function
G17	XY plane selection
G18	ZX plane selection
G19	YZ plane selection
G22	Subprogram call/figure rotation/user macro
G23	Subprogram return/figure rotation/user macro
G27	References point (original) collation
G28	Automatic reference point (origin) return I (memory-type reference point (origin) return)
G29	Automatic reference point (origin) return II (memory-type start position return)
G30	Automatic references point (origin) return II (memory-type second reference point (origin) return), third, fourth reference point (original) return
G31	Skip function
G34	Special fixed cycle (bolt hole circle)
G35	Special fixed cycle (line at angle)
G36	Special fixed cycle (arc)
G37	Special fixed cycle (grid)
G39	Corner switching
G40	Tool diameter compensation cancel
G41	Tool diameter compensation (left)
G42	Tool diameter compensation (right)
G43	Tool length compensation
G44	Tool length compensation cancel
G45	Tool position compensation (elongation)
G46	Tool position compensation (contraction)
G47	Tool position compensation (double elongation)
G48	Tool position compensation (double contraction)
G50	Scaling Off
G51	Scaling On
G52	Local coordinate system
G53	Work coordinate system offset (selection of the basic machine coordinate system)
G54	Work coordinate system offset (selection of the work coordinate system 1)
G55	Work coordinate system offset (selection of the work coordinate system 2)
G56	Work coordinate system offset (selection of the work coordinate System 3)

BIBLIOGRAPHY

Section Seventeen

Klein, Roy S., Manufacturing Technology for Designing Plastic Parts, Seminar Materials, Dearborn, MI: Society of Manufacturing Engineers, 1983.

Hill, John Jr., Engineering Thermoplastics, Business Opportunity Report P-01 5R, Stanford, CT: Business Communications Co., Inc., 1983.

Schwartz, Mel H., *Composite Materials Handbook*, New York: McGraw Hill Book Co., 1984.

Section Eighteen

Atkinson, William, "The Customer-responsive Manufacturing Organization," *Manufacturing Systems*, Vol. 8, Carol Stream, IL, Hitchcock Publishing Co., May 1990.

Axel, Johne, "Don't Let Your Customers Lead You Astray in Developing New Products," *European Management Journal*, Vol. 10, Tarrytown, NY, March 1992.

Cordero, Rene, "Managing for Speed to Avoid Product Obsolescence: A Survey of Techniques," *Journal of Product Innovation Management*, New York, NY, Elsevier Science Publishing Co., Inc.

Drozda, Thomas J., "Manufacturing Forum," *Manufacturing Engineering*, Dearborn, MI, Society of Manufacturing Engineers, April 1984.

Haskins, Robert and Petit, Thomas, "Strategies for Entrepreneurial Manufacturing," *Journal of Business Strategy*, New York, NY, Faulkner and Gray, Inc., Nov/Dec 1988.

Staff, Industrial Technology Institute, *Design for Manufacturing*, Ann Arbor, MI, ITI, 1988.

Keller, Erik L., "Two Technologies that May Reduce Time to Market," *Concurrent Engineering*, Vol. 1, New York, NY, Auerbach Publishing, July/August 1991.

Parker, Roysten V., "CAD/CAM Using I.D.S.," *SME Technical Paper*, MS78-971, Dearborn, MI, Society of Manufacturing Engineers, 1978.

SME Publications Staff, *Simultaneous Engineering, Integrating Manufacturing and Design*, Dearborn, MI, Society of Manufacturing Engineers, 1990.

Stalk, George Jr., "The Next Source of Competitive Advantage," *Quality Progress*, Vol. 22, Milwaukee, WI, 1989.

Tucker, Walter W. and Clark, Richard L., "From Drafter to CAD Operator: A case Study in Adaptation to the Automated Workplace," *SME Technical Paper*, MM84-629, Dearborn, MI, Society of Manufacturing Engineers, 1984.

Section Nineteen

Amrine, H.T., Ritchey, J.A., Moodie, C.L., and Kmec, J.F., *Manufacturing Organization and Management*, Sixth Edition. Englewood Cliffs, NJ: Prentice Hall, 1993.

Cubberly, W.H., and Bakerjian, R., Eds., *Tool and Manufacturing Engineers Handbook, Desk Edition*, Dearborn, MI: Society of Manufacturing Engineers, 1989.

Curtis, M.A.*Tool Design for Manufacturing*, New York: John Wiley and Sons, 1986.

Drozda, T.J., and Wick, C., Eds., *Tool and Manufacturing Engineers Handbook, Vol. 1*, Dearborn, MI: Society of Manufacturing Engineers, 1983.

Goetsch, D.L., *Advanced Manufacturing Technology*, Albany, NY: Delmar Publishers, 1990.

Machinability Data Center, *Machining Data Handbook*, Second Edition, Cincinnati, OH: Metcut Research Associates, 1972.

Wick, C., Benedict, J.T., and Veilleux, R.F., Eds., *Tool and Manufacturing Engineers Handbook, Volume 4,* Dearborn, MI: Society of Manufacturing Engineers, 1984.

Wick, C., and Veilleux, R.F., Eds., *Tool and Manufacturing Engineers Handbook, Volume 3,* Dearborn, MI: Society of Manufacturing Engineers, 1985.

Section Twenty

Veilleux, R.F. and Petro, L.W., *Tool and Manufacturing Engineers Handbook,* Vol. 5, Dearborn MI, Society of Manufacturing Engineers, 1988.

Juran, J.M. and Gryna, F.M., *Quality Control Handbook,* Fourth Edition, New York, NY, McGraw-Hill Book Company, 1988.

Section Twenty One

Cubberly, W. H. and Bakerjian, R., *Tool and Manufacturing Engineers Handbook Desk Edition,* Dearborn, MI: Society of Manufacturing Engineers, 1989.

Groover, M., *Automation, Production Systems, and Computer-Integrated Manufacturing,* Englewood Cliffs, NJ: Prentice-Hall Publishing Co., 1987.

Korth, H. F. and Silberschatz, A., *Database System Concepts, 2nd ed.,* New York: McGraw-Hill, Inc., 1991.

Vielleux, R. and Petro, L., *Tool and Manufacturing Engineers Handbook, Volume 5,* Dearborn, MI: Society of Manufacturing Engineers, 1988.